THE "GJÖA" EXPEDITION,
1903–1907.

THE

NORTH WEST PASSAGE

BY

ROALD AMUNDSEN.

The Gjoa's first meeting with Whalers after completing the North West Passage

ROALD AMUNDSEN'S

"THE NORTH WEST PASSAGE"

BEING THE RECORD OF A VOYAGE OF EXPLORATION OF THE SHIP "GJÖA" 1903–1907 BY ROALD AMUNDSEN WITH A SUPPLEMENT BY FIRST LIEUTENANT HANSEN VICE-COMMANDER OF THE EXPEDITION

WITH ABOUT ONE HUNDRED AND THIRTY-NINE ILLUSTRATIONS AND THREE MAPS

VOL. II

New York:

E. P. DUTTON AND COMPANY

1908

CONTENTS.

ILLUSTRATIONS.

Illustrations.

Illustrations.

CHART.

THE NORTH WEST PASSAGE

VOLUME II.

CHAPTER VIII.

PART II.

The Inhabitants at the Magnetic North Pole.

The seasons in these regions end just as abruptly as they set in. The Eskimo awakes one morning to find himself in the depth of winter, the sea is frozen over and the snow in places has formed drifts several yards deep. Now, there is no longer any excuse for delay in the building of snow-huts; they have all suffered enough from the cold during the past night. Soon the whole population of the colony is out selecting building sites. The main thing is to find a sheltered place, screened to some extent from the wind and not too far from water, as otherwise every drop they require would have to be procured at the cost of trouble and fatigue. The condition of the snow also is an important consideration; if it is not favourable the hut will not turn out a first-class job. The selection, therefore, of a good site for the hut is a very responsible task for the paterfamilias, and it often takes him a long time to decide on it. He carefully tests the snow with an instrument specially intended for the purpose, called a "hervon." This is a stick made of reindeer horn, straightened out like a long walking-stick. It is about four feet long. At one end there is a handle of reindeer bone and at the other a musk-ox bone ferrule (Fig. 3, p. 299, Vol. I). In the course of his examination he thrusts the "hervon" into the snow to "feel" its condition. It requires a very delicate sense of touch, developed by many years' practice and experience, to "feel" the condition of the snow. Anyone, by sticking a rod into the snow, can ascertain whether it is hard or soft, but to determine the number and condition of the various strata is a far more difficult task; for it very often happens that the snow drifts consist of layers swept together at different times and in different weathers, consequently they vary considerably in character. In one and the same snow drift you may find snow that has been beaten together into a compact mass by a storm, together with snow that has settled down on it gently in calm weather, forming a very loose layer, which is quite unsuitable for building purposes. Over this again you may get a hard stratum, and it needs the skill of an Eskimo to distinguish the loose layers in the mass of drifted snow. The ideal condition is attained when the drift has a loose layer of snow about one foot thick on the top, and a uniform mass of the requisite hardness below to a sufficient depth for making the blocks required; yet the snow must not be too brittle, as in that case the blocks are likely to crumble in course of preparation.

In order to obtain a correct idea as to how a hut should be built in the most approved style, we will pay a visit to the master-builder, Atikleura. He is standing just below the summit of the ridge beckoning to Nalungia to intimate that he has found a suitable spot and that she is to bring him his snow shovel. A glance at the site he has selected shows that Atikleura is a practical man as well as a man of taste. The position is well sheltered to the north, east, and west, and the crest of the ridge at the back will prove a barrier to the biting north wind. Towards the south the prospect is open and will have the full benefit of the sunshine. Close by there is a small lake or pond which will supply the most delicious drinking water for the family. The country hereabouts consists mainly of spacious plains and beautiful lakes. Meanwhile Nalungia has arrived with the snow shovel. This is made of a wooden board which Atikleura has obtained by barter from tribes dwelling further south, as there is no wood in Nechilli, nor does the smallest piece of drift wood ever find its way to these latitudes. The shovel is made in a very workmanlike manner, and excellently suited for its purpose as long as the snow is loose. For hard snow, of course, our iron spades would be preferable. It is strengthened at the lower end with reindeer bone. Now, the first thing that Atikleura does, is to shovel away the upper loose layer of snow, in the circumference within which he had planned to erect his hut. He does so with a true eye, as the large number of huts he has built in his lifetime has given him good practice. Then he draws out the knife which has hitherto been suspended by a loop on the bone peg at the back of his "anorak." It is quite a monster knife, enough to frighten anyone who had not seen it before. The blade is as large as that of an ordinary good sized butcher's knife and is made of iron, which has also come from the south; the handle is about a foot long, and is of wood or bone. Taking the handle with both hands he commences to cut out his ice blocks for building the hut. These are cut out to a size about eighteen inches wide, twenty-four inches long and four inches thick. If cut out in this way, the building site itself will yield sufficient material for the whole construction.

It is a pleasure to see how a good builder cuts each block so that it just fits where he sets it. Atikleura is a veritable prodigy at this work. Not one of his blocks ever breaks in pieces, although he appears to cut them out without any particular care. Just a cut here and there, then a kick, and the thin neat block stands separated from the mass of snow. All the blocks from Atikleura's hand are so exactly equal in size that they look as if they had been accurately measured. The hut is built up in spirals in the form of a haycock or bee-hive, so that one layer of blocks rests on the previous one and extends a little further inwards. In joining the blocks, the sides must be fitted to each other so that the walls are perfectly tight. The builder's skill can be gauged by the tightness of the hut; but even with Atikleura's skill it is impossible to avoid some few small chinks here and there. It is Nalungia's task to fill up these chinks. For this purpose she works the shovelled-up loose snow until it is as fine as grated sugar, for it is only when it is in this state that it can be used for making the joints tight. It is thrown up against the blocks as soon as they are placed in position and fills in every little hole and crevice. The walls of the hut rise quickly. As the blocks are cut out the ground is cleared downwards, and as they are set into their places, they serve to increase the height of the walls of the cleared site. Atikleura looks as if he had been standing on his head in a flour-tub; he is covered with snow all over; his clothes, hair, and beard are white as chalk. His long gloves prevent the snow from getting into the sleeves of the "anorak."

Building the roof of such a snow hut is a very complicated affair to the uninitiated. Many a snow-block did I get on my head when I essayed this work. The snow-blocks have to be set back gradually inwards, and when the work is nearing completion, the last blocks would appear to be literally suspended in the air, without any base or support. The last block (or keystone) which closes the roof, in the centre, is quite small, and in most cases triangular. To fix it in its position from the outside, it must first be juggled out through the hole which it is eventually to fill. This looks impossible, but the Eskimo achieves the impossible. With one hand he raises his block to the outside, through the hole at the top, and while holding it he cuts it into the shape of a wedge with the knife he holds in the other; and when he lowers it into the hole it fits it as if it had been moulded for the purpose.

Nalungia, aided by Errera, has perseveringly plastered over the outside of the hut with fine snow, so that it simply looks like a snow-heap. The outlines of the blocks are now quite concealed under the snow. But the hut is perfectly tight, as the fine snow works itself in wherever there is the slightest hole or crevice. The master-builder himself is not yet visible; he is still busy in the interior of the hut, where he is now completely built in. At last his long-bladed knife protrudes from the wall of snow, and with a rapid movement he cuts a hole just large enough for him to creep through. I am surprised to see how high up the wall he cuts the hole, as in all the huts I have hitherto seen, this entrance hole was quite down to the floor. Now Nalungia creeps in through the aperture, and I follow her to see what she is going to do in the way of further internal arrangements. I am at once enlightened as to why the aperture is made so high up; Atikleura has cut it on a level with the sleeping-berth, to expedite the work of "moving-in." He has constructed the sleeping-berth as follows:—He has first divided the hut by a row of snow-blocks into two compartments, of which the inner one is twice as large as the outer. He throws all the loose, refuse snow lying in the hut, into the inner compartment, until it reaches the level of the row of blocks, and there you have the "bedstead" quite ready. At the opposite end of the hut is another small erection, made of two blocks set on edge, and a third laid across them, like a table slab.

Now commences the moving in, through the aperture above the sleeping-berth. Large quantities of skins are thrown in and slung topsy-turvy upon the sleeping place. Next comes all the furniture—a drying grid, water bucket, cooking pot, blubber lamp, provisions, blubber, meat and fish, and lastly the women's personal belongings —which I dare not specify more fully. Now it looks as if all were over and Mrs. Nalungia casts an enquiring look at me, as much as to say, "Are you going to creep out?" I have no idea what is about to happen, but my curiosity prompts me to remain, thinking that anything much worse than I had seen before was hardly likely to occur; but I certainly was a little taken aback when the hole over the sleeping berth was suddenly blocked up again from outside and I was alone, with one lady, in a closed-up hut. However, as Nalungia did not seem to mind it in the least, why should I trouble? Disregarding me she set to work with a will. The heavy blubber lamp was first raised upon the little snow table near the wall opposite the sleeping berth. This lamp is made of a kind of stone they obtain from the Utkohikchyallik Eskimo; it is carved in the form of a crescent and is heavy and clumsy. It is placed upon three pieces of bone inserted in the snow slab, so that the inner edge of the crescent is turned towards the interior of the hut while the outer edge is towards the wall. The blubber bag is now brought out and a piece of frozen blubber taken from it; this is beaten with a specially made club of musk-ox bone until it is quite soft. Now she produces, from one of her repositories, a little tuft of moss which she carefully soaks with seal-oil—ugh! I remember with horror those mysterious "light pastilles"—and then she sets to work to get a light by rubbing pieces of wood together. The "pastille" soon sends out the most dazzling rays; the crushed blubber is put into the lamp, and a wick of moss is laid along the whole of the inner straight edge; this is sprinkled with seal-oil and ignited by means of the burning tuft of moss. The whole wick is now blazing and a brilliant flame lights up the roomy hut. I ask myself what in the world she wants with this brilliant flame, as she has now finished arranging the hut, and I am almost on the point of upbraiding her for this waste of precious oil, but I refrain, as I remember that an Eskimo never does anything without good reason. In fact it soon becomes apparent that here, too, my judgment is premature. Gradually an oppressive heat spreads from the mighty flame, and now I understand that her object is to cause the newly-built hut to settle well down at the joints. As the result of the heat thus produced, the snow blocks gradually close up till they may be said to form one single continuous wall.

While this is going on, Nalungia makes good use of her time, and gets the sleeping berth into proper order.

The waterproof kayak skins are laid next to the snow; these have been taken from the kayaks in the autumn, and will keep the moisture of the snow away from the reindeer skins neatly arranged over them, and the sleeping berth looks quite cosy. Again she turns her attention to the lamp and trims the wick—this has to be done frequently; the saucepan is then filled with snow and suspended over the flame by two cords, secured to two bones fastened into the wall. The family may want refreshment after this job. The drying grid, made of reindeer bone, strung over with a network of sinew thread, is now fixed up over the saucepan but not too near the fire. The skins will not bear too much heat. Finally, the "anauta," a small, round, thick, wooden stick with a handle, used for beating the snow off the clothes, is, by way of a finishing touch, driven into the wall. Everything is now ready. And none too soon; for at this moment Atikleura is calling from outside asking if he may come in. Nalungia casts a last critical look round the walls, and tells him to wait a little. He goes off muttering something which, translated, would sound very much like "d——d womenfolk" or something of the kind. Nalungia looks as though she meant to pay him out for his courtesy by keeping him waiting a little longer, and it is quite another half hour before she calls him in. Then an opening is made through the wall, right down to the floor, large enough for a man to creep through, and Atikleura's head appears through it. A moment later he is inside the hut; he takes off his soaking wet gloves and throws them towards his wife, who turns them inside out and hangs them on the drying grid; then she takes his coat, shakes it and well beats it with the "anauta," for it is important to remove every little grain of snow to prevent it melting and wetting the coat, which is then rolled up and thrown on the bed. The outer trousers are then treated in the same way and placed with the coat next the "anorak." Atikleura stands there in his under garb. This does not sound exactly "comme il faut" according to our ideas, but it calls for no comment among the Eskimo. He now walks up to the sleeping place and sits down, not, as we might do, on the edge, but well back so that he can rest his legs. Now the footgear must be removed, and this is not a very simple matter, as an Eskimo's footgear consists of five different articles. Outermost are the low reindeer-skin shoes, made with the hairy side inwards. For a man of Atikleura's high descent these are half-soled with sealskin. On the bottom of the sole there are some perceptible ridges which, on closer inspection, prove to be strips of skin sewn on to prevent the foot from slipping. Next come the "kamiks," which at this time of the year are exclusively of reindeer skin. There are two pairs of these. The outer are made of the hide from the reindeer's leg, which is short-haired and very strong. They are made with the hairy side inwards, and reach up to the knee, where they are laced up with a thong. Underneath these is another pair, exactly of the same length and appearance, but with the hairy side outwards. These are made out of the hide of a one-year-old reindeer, taken from the abdomen, as the skin there is very fine and soft. Between these two pairs of "kamiks" the Eskimo wears a pair of short reindeer-skin socks, with the hairy side outward;

NECHILLI ESKIMO IN THEIR SNOW HUT—COVERING THEIR FACES TO PREVENT BEING PHOTOGRAPHED.

and, lastly, another pair of socks next to the skin, with the hairy side inwards, so that altogether the feet have five different coverings. When I first saw this I thought that, after all, we were rather more hardy than the Eskimo, as we only used three articles of foot-gear; but on my first sleighing tour I realised that it was not simply for protection against cold that the Eskimo used all these articles, but, to a great extent, to protect the feet against the hard snow and ice on which they are always walking. With my triple foot-gear I became so footsore that I could scarcely walk. Like the gloves, all the foot-gear must be put up on the grid to dry. The inconvenience of skin clothing is that, unless kept well aired, it is very apt to absorb and retain any moisture. The Nechilli Eskimo did not know of sedge-grass; they put loose reindeer hair into their boots and take it out at night; this was better than nothing, but not nearly so good as our grass.

When Atikleura has removed his wet foot-gear, he puts on a pair of dry "kamiks" and a pair of low seal-skin shoes—"kamileitkun"—corresponding to our slippers. In winter these are used inside the hut only, but during the transition period between winter and spring they are worn outside. As far as the care of the outer man is concerned, Atikleura is now ready, and is therefore at liberty to think of the needs of the inner man. And these are not trivial, after the trying day's work. A fine salmon is served up, and all the members of the family partake freely. Frozen though it is, it seems to be highly relished, and very shortly there is nothing left but the clean-stripped skeleton. The saucepan, now full of fresh clean water—a few hundreds of reindeer hairs, of course, are not looked upon as impurities—is emptied, and refilled with snow and suspended again over the fire. Water is the only drink the Nechilli Eskimo know; no "half-and-half" of any kind is to be had there. They now announce that there is no more room in their stomachs for either salmon or water, and the meal is finished. It is time to turn in. Nalungia prepares the bed for the night, arranging the beautiful soft skins; Atikleura closes up the entrance securely with a block of snow, slips in under the large family bed rug, and there disrobes. Unlike the Greenland Eskimo, these people, of either sex, never disrobe in the presence of strangers, except in the greatest emergency. The guest of the family is assigned a place at one side of the hut—little Anni and Errera have turned in long ago—and the berth nearest the fireplace is reserved for Nalungia. She extinguishes the light and arranges her toilet in the dark. The large skin bed-rugs are their only covering at night. Vigorous snoring soon announces that they are asleep.

The scene outside is very different from the one which we pictured in the summer. The tents have all disappeared, and, in the peaceful moonlight, the low cupolas of the snow huts are almost merged in the snow-covered field. A stranger passing by would scarcely suspect that quite a little world is slumbering there, and, least of all, a world of glad and happy people, happier, perhaps, beneath their lowly snow roofs than many a rich and mighty one under a roof crowned with turrets and battlements. Rancour and envy, calumny and malice, are banished from the world of this ice desert; the peace of the night is unbroken, and the moonlit atmosphere is pure around the abodes of these men.

Chapter VIII.

There are still a great many trifles to keep the inmates busy in the hut next day; in the hurry, in which all was done yesterday, it was impossible to get everything arranged as it ought to be, so that there is still a good deal to be set right or improved. Nalungia's first thought is to have a window in the hut. True, a snow hut, even without a window, is light enough to enable them to see to work by day, but with a window it will certainly be much lighter and more cheerful, and they will also be able to see and judge of the weather without going out. Atikleura, who is by no means heedless of the wishes of his wife, goes down to the ice-bound lake and cuts out a suitable oblong slab for a window, which he puts into the wall over the entrance door, the most elegant window imaginable.

Time passes, and the moon, of high importance to the Eskimo, soon reaches that particular position in the heavens that permits the women to resume their sewing. This is a busy time. There are skins to be dressed, cut out, and sewn. To see an Eskimo woman cut out garments is most amusing. She has no chalk for marking out, but she has strong teeth. The skin is folded to the required shape, a mark is bitten in with the teeth, and the garment is then cut out with the "olo." Many do not even trouble to bite marks, but cut out by guesswork, with the sure eye acquired by years of experience. When the Eskimo woman sews for her family or herself, the stitching is done in an exemplary manner and the stitches are fine and small;

but if she does sewing "to order," for a "Kabluna," then it is execrable, long coarse stitches that will not hold for a day. Consequently, when we want serviceable clothes, we buy them second hand from the wearer.

It is an art to make skin clothes so that they are comfortable, both as regards cut and make-up. The Eskimo sew the skins together edge to edge, thereby avoiding the thick seams produced by overlapping. On board the "Gjöa" we had skin clothing made in Norway and Siberia, but we should have been crippled in a single day, had we worn these clothes with the seams next to us, they are so thick and coarse. On the other hand, I have repeatedly gone about in Eskimo clothes in warm summer weather, with the hairy side out and the seams next to me, without the least discomfort.

At the end of October the Eskimo appear in their new clothes. The "swells," Ahiva and Oyara and their wives, always lead the fashion. A brand-new Eskimo dress is really very attractive. They wear two tunics or "anoraks," one with the hairy side inwards, and one reversed. In design they are very much like a dress coat. I do not know with whom the design originated. Among the Nechilli Eskimo the tails of the "anorak" are not very long, they scarcely reach below the knee; but some other tribes wear them reaching right down to the heels. The outer "anorak" is elaborately trimmed, and is made of a thicker skin than the inner one. Both garments hang loosely on the figure, so that the air can circulate freely. They also wear two pairs of trousers, one with the hair outwards, and the other with the hair inwards. The outside trousers are often decorated with trimmings, while the inside ones, of course, are plain. They are tight round the waist, but loose at the knee.

AHIVA AND ALO-ALO IN HUNTING GARB.

Both the "anoraks" and the trousers are often edged with fringe.

Before the members of the expedition had become quite accustomed to the Eskimo dress and had adopted it, many of us thought it ridiculous for grown-up menfolk like ourselves to go about wearing fringe to our clothes, so we cut it off. But I had my scruples about this, as I had already had evidence that nothing either in the Eskimo's clothing or other arrangements was, in fact, without meaning or purpose, so I kept my fringe and put up with the ridicule. However, he laughs best who laughs last; one fine day the "anoraks" from which the fringes had been cut off, commenced to curl up, and if the fringe had not been put on again pretty quickly, they would soon have looked like neckties.

There are merry doings in Nechilli at Christmas time. Although they have no idea of our Christmas or our reason for celebrating it, they have their winter festival, fully corresponding to our Christmas just about that time. The food depôts are full of fish and reindeer, and the days are given up to eating, drinking, and amusement. They have built a large igloo, which serves as a common assembly room and entertainment hall. Some of these public igloos are quite palatial and will hold over fifty people. The amusements consist of gymnastics, conjuring, singing, and dancing. Gymnastics are cultivated by men of all ages; even old Kachkochnelli takes his part, and acquits himself as well as the younger men. Having no horizontal bar they had improvised one with the means at their command; a long cable was formed by five sealskin straps laid together, and a second sealskin strap coiled tightly round it: this constituted a very reliable rope. Now the question was, how to get it fixed, and that was not so easy, as a snow wall does not really afford any strong hold. But the Eskimo know how to help themselves. They drill holes in each of the two opposite walls, pass the two ends of the rope out through these, and fasten them to two wooden bars secured in the snow on each side of the hut. Now this makes a capital elastic "horizontal bar," and the display commences. I was dumbfounded to see these people perform many of the gymnastic feats I remembered from my boyhood, and they really did them very gracefully. They were supple and agile. I was tempted to display some of my former agility, and show them what I had been capable of, but I came to grief. It was of little use for me to excuse myself on the ground that I was unaccustomed to a rope in place of the familiar horizontal bar; the failure of my intended exhibition aroused general hilarity among all present, both Eskimo and Kabluna.

For their conjuring tricks, they do not need this large hut; they can be performed anywhere. These performances, as a rule, have some special object: to drive away sickness or to ensure a good catch, etc. In spite of my persevering investigations, I was never able to find out what the qualifications for a conjurer or "angekok" are. There are various grades, some high, some lower class, and some quite inferior. Kagoptinner thus was a very great magician, in fact, as I have said before, the greatest of the tribe. Old "Praederik" was also one of the leaders, but not so great as Kagoptinner. We were never permitted to be present at these performances, but once I succeeded, by pure accident, in getting a tolerably good insight into the affair. I was calling on the "Owl" for a chat. Outside his hut there were two Eskimo, who addressed me as I was going in; I understood it was something particular they wanted me for, but it was not until later that I found out what it was. On my way there I had heard loud shouting; I took it for singing, of a more weird description than usual, and continued my walk towards the hut. I remained at the inner entrance to the hut, which was so low that I had to creep through it, and lie down on all fours, to see what was going on. I soon saw that old "Praederik" and his wife, a horrid old woman, one of the few I could never get on with, were practising "sorcery." The sleeping place on which they were, was almost in darkness. As far as I could see by the dim reflection of the little "light pastille" (a tuft of moss impregnated with seal oil), which was the only light in the hut, I thought I could faintly discern the outlines of the two individuals. The "Owl" and Umiktuallu, with their families, were spectators. They stood as far as possible away from the performers, and all looked very solemn. Luckily I was not observed at first, and was able to watch them all for a time unnoticed. The old woman was shrieking outrageously; the yells of old "Praederik," which under ordinary circumstances might have been considered a very creditable performance in this line, were quite drowned by hers. I could not see what else there was going on on the bed-place, and a movement on my part, to get nearer the performers, led to my being discovered. However, quite unconcerned, I raised myself quietly and wished them good evening. But I had far better have abstained from doing so, for the old dame now uttered such a terrific howl as to make me fly precipitately. I have always had the greatest horror of women in that state. However, a very few minutes after, the "Owl" came out and told me that the performance was over, old "Praederik" having now "run himself through with a spear." I did not consider this very pleasant, as I quite imagined him to be bady injured and dying, but the only answer made by the "Owl" to my question as to his condition, was an invitation to come inside the hut. There the old rascal was sitting on the sleeping place, apparently in the best of health and humming softly. His wife did not appear to have quite recovered from her frenzy, she was swinging her arms about and scowling at me. I did not then venture any allusion to what had been going on, but indulged in a little general conversation, and went on board. Later that night, old "Praederik" came and showed me two holes in his "anorak, one behind and one in front, as irrefutable proofs of his having transfixed himself with his spear! He was a very decent, honest fellow and I feel confident that he really imagined he had. Nor was it to be wondered at, that the mysterious howls uttered by the better half had for a time deprived him of his sound senses.

The Eskimo are not altogether without forethought

PRAEDERIK AND WIFE IN THEIR SNOW HUT.

for the future. Their stock of meat and fish will last over Christmas and a little way into the new year. According to their law, seal-catching must not commence before the middle of January, and even then it is carried on only on a small scale for some time, as the seals, which have very sharp ears, can hear the huntsman's step a long way off, while the layer of snow on the ground is thin, and consequently they can keep out of his way. Therefore, from the middle of January up to some time in February is their period of greatest privation. Thus it happened during their stay at Gjöahavn in 1905 that, for several days, just about this period, some of them were absolutely starving. Nor could we help them much, for if we gave anything to one they would all come, and we should soon have had the whole colony to feed; so I had to lay down the law that no food was to be distributed—and enforce it.

One night, after seal-catching had commenced, I was invited to witness the "kėlaudi," the Eskimo's favourite festival, held in order to propitiate the higher powers to induce them to favour a good catch. The air was clear and frosty, and the vast silent desert was lit up by the moon so brightly that one could easily have read by her rays. In the midst of the many igloos of the camp, the large gala igloo, erected for the occasion, towered above the rest, with bright light streaming invitingly from all its ice-windows. We arrived early, so as not to miss anything. The hut looked very well inside, being brilliantly illuminated with "light pastilles." A large ring of snow-blocks had been set up in the centre of the floor. Some of the men of the colony had already arrived, and entertained us as well as they could. They were decked to-night in their lightest and most elegantly-ornamented reindeer clothing. Some of these were actual masterpieces of taste and skill. By and by the rest of the audience arrived—Anana, Kabloka, Onaller, Alerpa, Alo-Alo, and whatever names the others are called by, not to forget "Nalungia," there being at least ten of that name. "Örna" (the she-eagle) was the last of a row of at least twenty women, who all sat down silently and demurely on the snow-blocks arranged in a circle. They certainly did not look as if they were in a festive mood, any of them. The men took up their places at random around the women, and there was soon a full muster of them. In contrast to the women, they were all lively and full of fun and laughter. It looked as if they were the only ones who were to enjoy themselves. At last the leading "senior" appears. To-night it is Kachkochnelli who acts in this capacity. He is arrayed in a light embroidered reindeer-skin dress, but he is wearing a cap and gloves. He brings with him the precious "kelaudi," the musical instrument of the tribe, which consists of a hoop of wood like a barrel hoop, covered with thin tanned reindeer skin, and fitted with a handle; the drum-stick is a small club of wood, covered with sealskin.

The entertainment now begins. Kachkochnelli enters the ring; thereupon Anana lifted up her voice and started something which I suppose I must call singing, though I find it rather hard to use the word in this connection, and the other women joined in. I have never heard anything so monotonous, its effect is still worse when chanted in chorus. But there must be some sort of fixed tune in this chanting of four notes, because they all manage to keep together. As the other women join in, Kachkochnelli commences to dance and beat the drum. It was not exactly a graceful dance. Keeping in one spot, he raises first one, then the other leg, and sways his body forward and backward, uttering loud yells. All the time he vigorously belabours the drum with his drumstick, striking it not on the skin, but on the frame. The result of all these efforts is a deafening din. Kachkochnelli's dance gradually becomes less and less energetic, and after about twenty minutes he stops. The women's chant, which has been keeping time with the dancer's movements, dies away simultaneously with the cessation of the dance. Then the next man enters the ring. There does not appear to be any order of precedence among the Nechilli, whoever happens to sit nearest, and is willing to perform, comes up unceremoniously into the ring, and the same dance, the same yells, and the same chant are repeated, without a shade of variation. But I noticed that the women took turns in leading the singing. When Kirnir, an Ichyuachtorvik Eskimo was dancing, it was a woman of his tribe who acted as precentor, and when Nulieu, the Ogluli Eskimo performed, an old cross-eyed Ogluli woman led the singing. It also seemed to me as if the tune varied slightly for the various tribes, but I should not like to be certain on this point. As I have already hinted, I have not a good ear for music.

I have seen this dance and chant described in several books of travels, and all the authors are unanimous in declaring that the performers worked themselves into a state of frenzy. This I cannot endorse. According to my very careful observation they were all quite normal and in their full senses during the whole dance, even when it was at its height. From the descriptions, I had expected something far wilder, and was therefore disappointed. It is altogether incomprehensible to me in what the pleasure of this performance consists. The performers looked bored, particularly the poor women, who had to repeat the same notes *ad infinitum*. In fact, they seemed quite delighted when there were no more volunteers, and immediately disappeared from the hut. This performance lasted about three hours, and had I known that it all consisted in a repetition of the first "turn," I should have come away much earlier.

These dances were performed throughout the winter. Frequently even after a fatiguing day at seal-hunting, after ten hours' toil on the ice in storm and cold, they would proceed direct to the dancing igloo for this mad exercise. The children, particularly the little girls, also had their own amusements of this sort. Two little girls, standing face to face, raise their shoulders, bending forward towards each other, and wriggle their bodies about, rapidly uttering a great many unintelligible sounds through their noses. All this is done with the profoundest gravity. Again they squat down opposite each other, with their knees tucked up to their chin, and hop up and down, mumbling some formula of words, with the same profound gravity of countenance. Though the pleasure of this game is, perhaps, not very great, to judge from their expression, it is at any rate fairly good gymnastic exercise. They also had a number of other games, but did not seem to care much about them.

As I have said before, seal-catching does not really begin nor assume any importance until some time in February, when the snow falls heavily and accumulates in drifts, yards deep on the ice, so that the seal cannot hear the step of the huntsman. Then the Eskimo's time of privation is over, the empty larder will soon be replenished. Seal catching and reindeer hunting constitute the staple industry of the Nechilli; and as their methods of seal catching are almost unknown to the outer world, I will attempt to describe them as well as I can from the observations I made during the excursions I took in their company.

It is a raw, dark morning in February. A gale is blowing from the north-west, and snow falling so thickly that we look as though we had emerged from a flour bin. It is scarcely 8 A.M., but there are lights in the igloos and the whole camp is astir. There is every indication that seal catching is about to begin, yet it is difficult for a stranger to understand why they should propose to go out in this awful weather, when yesterday they were walking and loitering about in the bright sunshine, with a dead-calm atmosphere. But the Eskimo's plans and calculations are always a mystery to us ; he is governed by his own laws, known to himself and to him alone. However, they always have a reason.

OGLULI ESKIMO REPAIRING HIS SLEDGE.

It is nearly 9 A.M. before they have all finished dawdling, for the Eskimo can dawdle, and that to perfection. To-day they are all starting out together in one party ; at other times they generally go out in small detachments. They are not all equally well equipped. Kachkochnelli is a man of method and order, who always keeps his outfit in perfect condition, so that, by studying him, we get a correct idea of what a proper seal-catcher's outfit should be. The first thing we notice is that he has his skin clothing closely laced up all over, so as to be impervious to the air. He finds the open clothing, used at other times, too cold for a day like this. Suspended from a button on his back is his indispensable snow knife, partly covered by the hunting bag, the "tuttirea," which hangs above it on a sealskin strap, passed over the shoulders and across the chest. The hunting bag contains the following implements (see illustration of Eskimo implements, p. 299, Vol. I): a harpoon, "helmiaki" (Figs. 13 and 14) with harpoon line, "togakchyea" (Figs. 13 and 14); two appliances for observing the seal with, called "illa" (Figs. 15 and 16) and "kiviuchyervi" (Figs. 17 and 18); a hole protector called "anokchyleoiritkun" (Figs. 20 and 21); two small wooden pegs which serve as rests for the seal spear called "na-a-makta" (Figs. 11 and 12) and a strap for hauling the seal up called "okchyeun," besides some small pins of reindeer horn to stitch up the incision made in the seal's body, called "topota" (Figs. 7 and 8). This is the whole contents of the bag. The bag itself is square, mostly of reindeer skin, though frequently, for want of the latter, they are also made of the skin of the Arctic fox. In one hand he carries the remaining implements and weapons, the seal spear called "onaki" (Fig. 10), the hole-finders or probes (Fig. 10) called "hervon" (Fig. 3), the hole examiners called "hervatavra" (Fig. 6), and a spoon called "ilaun" (Fig. 9). In former times, and even until a few years ago, all these weapons and implements were made exclusively of reindeer horn; now some of the parts are often made of iron.

With the other hand he leads his dog by a reindeer skin strap. However, they do not all take dogs. There are about forty men in the party to-day, ranging in age from fifteen to fifty. They have a great deal to talk about. One would think they were living in a world full of stirring events, and offering a variety of topics for conversation and discussion, and not here in an ice field, which had been lying silent and desolate for æons, and where life from day to day, yea, from century to century, has gone on in changeless monotony. They proceed in a body over the ridge, but as they approach the ice they deploy in skirmishing order, gradually extending the intervals as they advance, until after a little time their line of march covers a considerable distance. Kachkochnelli presses ahead, humming softly and talking to his dog. There is nothing remarkable about the dog, neither a particularly high-raised head nor intelligent eyes. Just "a wretched cur" would perhaps be the best description of him. But "you must not judge a dog by his coat." Wretched as he is in appearance, I do not suppose his master would exchange him for the finest pointer, Gordon setter, or whatever else all those fine thoroughbred dogs may be called; for he has this merit, which renders him indispensable in these regions, that he knows how to track the seal. Suddenly he darts out on one side, stops and searches the snow carefully, and then lies down flat, leaving the digging operations to Kachkochnelli, who at once probes about in the snow with the "hole-finder," the same staff he used when testing the

YOUNG NECHILLI ARCHERS.

snow, with a view to building his igloo. Apparently the very first probings are satisfactory, for he at once slips the strap of his hunting bag over his head, takes the knife from its button, and with it proceeds to remove the layer of snow, covering the hole he has found in the ice. But this is not done without a previous thorough examination (the seal has many holes besides the one he resorts to for breathing), to see whether the hole he has found is really a "breathing hole" still in use, or only an old abandoned hole. Kachkochnelli lies down flat on his stomach, in the snow, and smells the hole. His keen sense of smell never deceives him. To-day, fortune favours him; he has

NECHILLI ESKIMO EQUIPPED FOR SEAL FISHING.

struck a genuine "breathing hole," evidently frequently resorted to by the seal. By a loud shout he intimates to his nearest comrades that he has a "find." This shout does not disturb the seal, yet the hunter takes care to move about the snow with the utmost caution, for he knows the seal is exceedingly timid of any sound of steps on the ice. The holes which the seals keep open during the winter are not large at the upper surface of the ice, but only just large enough to enable the seal to put its snout through to breathe. The hole gradually enlarges downwards, attaining its largest dimensions at the lower surface of the ice. Now, the first thing the hunter does, after having made sure he is on the right scent, is to cover up the hole carefully with snow, so that the seal may not scent danger, should he visit the hole while preparations are in progress. A portion of the layer of snow lying over the hole is removed by the hunter, who then digs with his spoon through the remaining snow, working his way right down towards the hole and throwing the dug-up snow away to the side. When he has got to a sufficient depth, he thrusts his "hole examiner" into the hole, in order to examine its interior more closely. Having done so to his satisfaction, he plants the "hole finder" firmly in the hole with one hand, and by shovelling snow round it with the other, fills up the excavation. The hole-finder being now carefully withdrawn, there is a narrow passage left in the snow, reaching right down to the seal's breathing hole, a passage of about the size of a halfpenny, yet large enough for his observations. If it is early in the year, and the layer of snow is thin, the hunter will immediately detect the approach of the seal without any aids other than his natural senses. But at this time of the year, when the snow lies so deep, he must have recourse to his ingenious devices. The appliances used for this purpose are the "illa" and the "kiviuchyervi" already referred to;

either may be used. Kachkochnelli prefers the "illa," and he also shows me how the "kiviuchyervi" is used. This latter is made of thick reindeer sinews, and is not unlike a very small grapnel with two claws. To this there is always attached, by a cord, a bunch of swan's-down, securely fastened on, so that it may not get loose. With extreme care he first pulls out a single piece of down from the bunch, and attaches the ends of this, with a little saliva, to the claws of the grappling-hook, so that the swan's-down forms a curved line between them. This is then lowered into the hole till the little cross-piece rests on the surface of the snow. As the whole appliance is not more than two inches long, it is easy to see the extended swan's-down in the hole below. If the wind is very rough, so that the wall of snow thrown up does not fully protect the hole from drifting snow, the "hole protector" is set over it. This looks something like a very small candle shade, and is made of transparent sealskin. Through this the hunter can watch the swan's-down. But I have never seen this cover in use. As soon as the seal comes within several yards of the hole the agitation of the water extends to the hole and sets the delicate swan's-down in motion. But the hunter's sharp ear and sure judgment alone can determine the right moment when to strike with the spear through the hole.

The "illa" affords a surer and more distinct indication of the seal's arrival at the hole, and the majority of the Eskimo therefore prefer it to the "kiviuchyervi."

This second appliance consists of two exceedingly thin rods made of reindeer hair, connected by a thin cord. One of these rods is eight inches long, the other is about twenty inches in length, and has a thin disc of bone about as large as a sixpence at its lower end. The cord joining these rods is fastened through

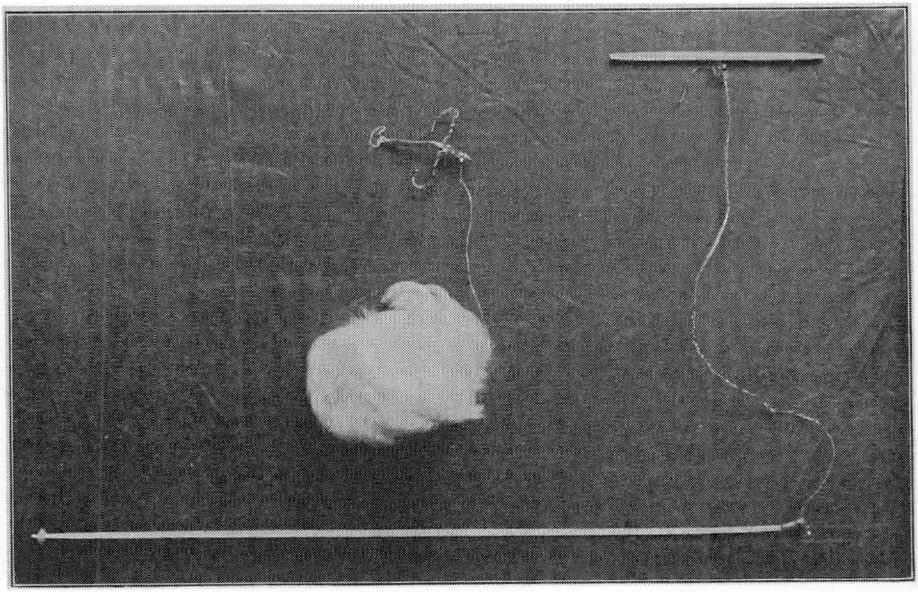

THE "ILLA" AND THE "KIVIUCHYERVI."

a hole at the top of each rod, and is about ten inches long. The "illa" is also more easy to fit up in the hole, than the "kiviuchyervi," as swans-down is not a material easy to handle in a storm and at a temperature of $-58°$ Fahr. To fit up the "illa" in the proper position is quite a simple matter; the shorter rod is stuck into the snow by the side of the hole, the longer one is lowered into the hole itself. When the little bone disc at the end of this rod comes down on the broken thin ice that covers the water in the hole, it rests on this ice, and thus supports the "illa" above the water, and the thread which connects the two rods remains lying quite slack on the snow.

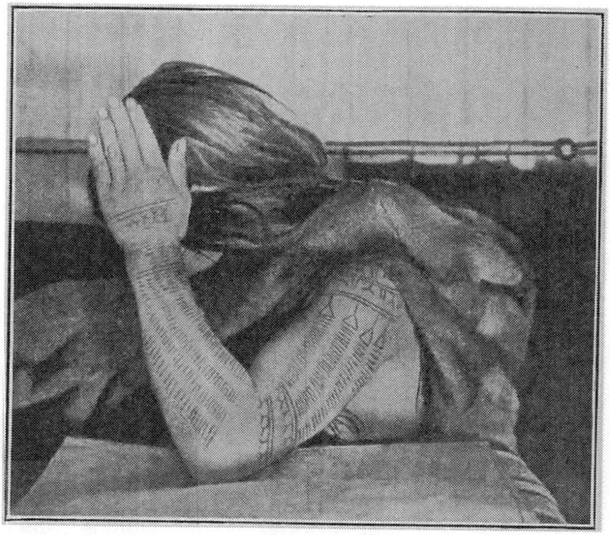

TATTOOED ARM (UTKOHIKCHYALLIK ESKIMO WOMAN).

During all these preparations the hunter has had the hunting bag lying under his feet as a mat, partly for warmth and partly to muffle the sound on the snow, lest it should scare the seal away. When the "kiviuchyervi" or the "illa" is in position over the hole, the seal spear is got ready. This consists of three parts, of which the middle one or shaft is about forty inches long by one inch thick, and made of wood or bone. One end of this shaft has attached to it a round projection, about two feet long, now mostly made of iron, though spears with reindeer horn projections are still used. But iron is greatly preferred to reindeer horn as the whole spear is more rigid and affords a surer thrust. This iron or reindeer horn projection terminates in a point which

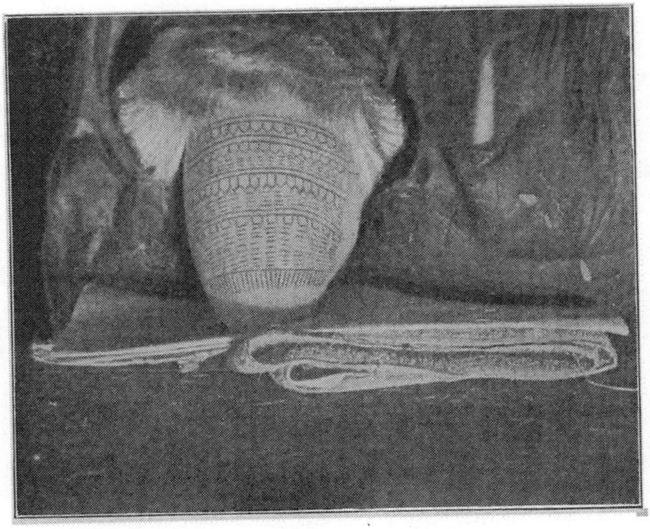

TATTOOED THIGH (NECHILLI ESKIMO WOMAN).

fits into a recess in the harpoon. To the other end of the shaft is attached the third part of the spear, which serves to enlarge the hole after the seal has been caught. This part may also be of iron or horn, but is most frequently of horn though iron is preferred. This is lashed to the shaft, and is curved and pointed at the end.

The harpoon is now attached to the spear, and the

harpoon line coiled up ready for running out, and tucked under a thin cord of reindeer sinew, extending the whole length of the spear shaft. This cord is fastened at both ends of the shaft. The harpoon is about three inches long and is now usually made of iron or copper. Only the extreme point of the harpoon is sharpened, so that the incision made in the skin of the seal may be as small as possible. Close to the point there are two barbs, curved back a little, so as to prevent the harpoon from slipping out again when once it has penetrated. As soon as the harpoon has entered the seal's body, it is torn away from the spear, and all the strain is brought to bear on the harpoon line, which is securely fixed through a hole in the centre of the harpoon, so that the latter will tilt over to a transverse position as soon as there is any tension on the line. The harpoon line consists of plaited reindeer sinews, and may be round or flat, but the crack hunter always prefers it flat. If flat, it is five-eighths of an inch wide and about thirteen feet long. It is always made of specially selected strong sinews.

When the whole spear is ready and in order, it is deposited on the two small wooden rests, "na-a-makta," fixed in the snow. The Eskimo now pulls off his long gloves, slips the left arm out of his coat-sleeve and places it across his chest underneath the coat, at the same time pushing the right hand into the empty left sleeve. Experience has taught him that this is the best way to keep his hands warm and yet be ready for action at any moment. The gloves are meanwhile held between the legs. Then he bends his knees a little and leans forward, keeping his eyes riveted on the hole. Everything seems in his favour to-day. The slight oscillations of the "illa" show that the seal is approaching. Owing to the rocking motion imparted by the agitation of the water to the thin, broken ice on which it rests, the "illa" bobs up and down. The Eskimo draws himself up, and in a moment

THE "OWL" TROUT FISHING.

both hands are back in position. With the right hand he grasps the spear, with the left he seizes the coiled line, and with every muscle tense, he prepares to strike. As the seal puts his snout up into the breathing hole, he pushes aside the fragments of thin ice, and the "illa" slips off and sinks, until checked by the cord which connects it to the other rod fixed in the snow. This is the signal. Concentrating all his strength, the Eskimo throws his spear. With unerring aim it darts down through the narrow passage and into the hole in the ice. It has caught! Quick as lightning he withdraws the shaft, sticks it into the snow by his side, and lets go the coiled line held by a running loop round his left hand. The seal runs out the line to its full length. But, after having been under water for some time, and being short of breath, its strength soon begins to fail. The Eskimo perceives this, takes the running loop from his hand and slips it over his foot, so as to have both hands free. He now hacks out the hole to a sufficient width to permit of hauling the seal up through it, and then pulls in the line. If the seal is not already dead, the iron mount of the spear shaft is thrust through its eye into the brain, and any resistance is prevented.

The strain of the rapid movements has hitherto been sufficient to keep the bare hands warm. But now the hunter puts on his gloves again. With the curved pike at the other end of the spear he pierces a hole through the upper lip and jaw of the seal, draws a strap through, and hauls it up. The harpoon is drawn out, and the hunter's work is done. Meanwhile, his nearest comrades have come up and assisted in the hauling, if required. With his knife he now makes a little incision in the seal's belly, and removes the liver and kidneys, and all present regale themselves on these dainties and a little blubber. On a day like this, when the temperature is so low, the flow of blood can be staunched by filling the incision in the seal with a mixture of snow and ice, which freezes instantly and closes up the aperture. In the spring, when the temperature is higher, the incision is stitched up with the reindeer horn pins ("topota"). When the seal is ready for removal, all the dogs at hand are harnessed to it by straps, and the hunters return home with their booty.

On the way home a Polar bear is sighted upon a hummock; he is away from the wind, so that our dogs do not get scent of him before he is sighted. Then they break away, all five of them, rush madly at the bear, and a battle royal follows. Bruin is quick when on his hind feet, and quick on all fours; he claws and strikes and snaps, but the dogs dodge him by quick turning, and he rarely gets near them. However, they do their duty, and detain him by worrying him until the Eskimo come up with levelled spears. Now the fight assumes a more serious character. To attack a bear at close quarters requires a brave man's courage, and the numerous scars and wounds the Eskimo bear on their bodies show plainly enough that they do not always get off scatheless in these hunts. At last Umiktuallu gives Bruin the *coup de grâce*, and he collapses. The flesh is divided among all the hunters, but Umiktuallu retains the skin.

On their return home the booty is handed over to the housewives. The seal is not too large to be hauled into the hut and skinned there. Nuyakke is a skilful woman, and the seal is soon skinned and quartered. The flesh and blubber are shared out equally among all; nothing is thrown away; the skin and entrails belong to the one who has killed the animal. The entrails, which to these people constitute the *bonne bouche*, are cleaned out and put into the pot to warm; it would be wrong to call it "boiling." This is a happy time in the Eskimo hut, though the hut itself looks nothing more than a mere snow drift. Kachkochnelli lies extended at full length on the sleeping berth, and hums or sings merrily, for rest is delightful after a long and fatiguing hunt in storm and cold. Nuyakke is busy with the cooking pot, and Kallo and little Nulieiu stand by with longing eyes. There is plenty of blubber, so there is no need to be sparing of the flame. The light also reveals some little faces in the doorway, but Nuyakke is a prudent housewife, who knows that if she gives to all who beg, there will be nothing left, and she pretends not to notice the intruders. But, of course, she must cut off a few inches of "tripe" for Kallo and Nulieiu.

The seal met with in the Nechilli's hunting field is exclusively the small species of seal; it is very fat, and its flesh is delicious. The reason why the larger species of seal never find their way there is that the water is rather shallow. The tract right across Matty Island to Boothia Felix and the Eta Sounds is too shallow for the large species of seal. But the latter are to be found plentifully enough on either side. In Ogluli the Eskimo catch them in large numbers. Strangely enough, these large and very powerful animals are caught in the same way and with the same appliances as the small seal. It seems incredible that one man should be able to hold one of these large animals on a strap. But it can be done at a pinch. Tolimao, a man about 5 feet 10 inches, strong and thick set, ranks foremost among these seal catchers. Last winter he harpooned a huge specimen of the large seal. It was tough work, but Tolimao had never let go before, and he did not mean to do so now. He dug his heels firmly into the snow, threw himself back, and held fast with both hands. The seal was too strong for him, pulled him over, and dragged him down into the hole, arms, head, and shoulders (of course, the breathing hole of the large seal is considerably larger than that of the small species); he did not slip in farther, but remained thus lying in a heap right across the hole until his comrades released him, but he never let go the line, and the huge seal was his.

When the spring is at hand towards the end of March, the time arrives when the seals bring forth their young. Lest they should be overtaken unawares by the interesting event when under four yards of ice, they commence to enlarge one of their numerous breathing holes in good time. They also dig their way little by little into the snow above the ice, until they eventually make an excellent snow hut, with the ice for a floor, and the mighty bed of snow above it for a roof. Here the young are born. They are often scented by the dogs; if the cubs are large enough they will plunge into the sea with the mother; if they are newly born or too young they fall a prey to the Eskimo.

Thus time passes away. June comes, and the snow

melts away from the ice. Then the seal comes up upon the ice to bask in the sun, and enjoy the sight of the open sky and the clear day after months spent in the gloomy deep. At this period the Eskimo catch large numbers of seals. They espy them as dark specks far out on the ice, and steal up to them; to reach them is an exceedingly difficult achievement. The hunter is armed with spear, knife, spoon, and accessories, the "topota." He, moreover, throws a small sealskin over his shoulder, and goes out towards the dark speck as far as he dare. Then he lies down and crawls. The seal is just as wary upon the ice as he is under it, and if there are no hummocks to serve the huntsman as a screen his task is no easy one. With his eyes riveted on his quarry, he wriggles forward; if the seal raises his head he must stop and lie flat on the ice till the seal has become reassured. When he has got quite near he places the sealskin under his elbow and glides forward on this, to deaden the sound. If the seal shows signs of alarm, the hunter tries to imitate its congeners by grunting and scratching the ice with his spoon, thus producing a sound like that made by the flippers of the seal scraping the ice. In this way he steals up to the animal, and, if he has been fortunate enough to avoid scaring it till he is within throwing range, he suddenly rises and throws his spear with a strong hand and a sure aim. Of course, it is at this critical moment that many seals escape, but a good many fall a prey to the hunter. Our Nordland seal-catchers can tell many a tale of hardships undergone in seal hunts on the ice. Frequently several consecutive days and nights are spent in seal-hunting, and the seal-hunters carry provisions with them and take their meals and camp over-night on the ice. They are hardy fellows.

July comes round once more, the sky is blue, the sun shines warm, and flowers abound on the hill-sides. In Nechilli tent after tent has arisen, and the kayaks are ready for launching.

I have never been able to find out exactly how the various tribes divided their hunting-grounds. But I believe I am not far out in stating that the Ichyuachtorvik Eskimo have their seal hunting-ground from Matty Island northward, the Nechilli from Matty Island southwards to Ogle Point on the mainland, and the Ogluli from Ogle west through Simpson Strait and out into Ogluli. Thus the Kilnermium Eskimo retain a very large field, from the Coppermine River to midway into Ogluli; perhaps this tribe is more numerous than the rest, or possibly this region is less rich in seals, but I should scarcely think so. Very frequently it happens that two tribes meet while out hunting. Such an encounter, far from leading to strife and bloodshed, is the signal for a round of festivities. Therefore the boundaries between their respective hunting-grounds can scarcely be very strictly drawn.

As regards computation of time, the Eskimo keeps strict count within the space of the current year. If he is to reckon by years (or, as he puts it, "summers and winters") he gets sadly confused and arrives at the strangest results. Old Kachkochnelli was once asked to tell us the age of his daughter Alerpa or Kodleo. He puzzled over it for a long time; in the usual Eskimo fashion he counted on his fingers, and paused now on the middle-finger now on the thumb, with a deeply pondering mien, At last he has solved this arithmetical problem; Alerpa, an adult, a fully developed woman, according to his calculation, was seven years of age. But he kept the most accurate count of the various months of the year. We could, for instance, make an appointment for a certain time many months ahead, and the appointment was always kept.

According to the "Owl's" statements, the Eskimo divides the year into thirteen months, called as follows:—

1. *Kapidra* (January) signifies: "It is cold, the Eskimo is freezing."
2. *Hikkernaun* (February), "The sun is returning."
3. *Ikiakparui* (March), "The sun is ascending."
4. *Avonivi* (April), "The seal brings forth her young."
5. *Nechyialervi* (May), "The young seals are taking to the sea."
6. *Kavaruvi* (June), "The seals are shedding their coats."
7. *Noerui* (July I), "The reindeer bring forth their young."

8. *Ichyavi I* (July II), "The birds are brooding."
9. *Ichyavi II* (August), "The young birds are hatched."
10. *Amerairui I* (September), "The reindeer is migrating south."
11. *Amerairui II* (October).
12. *Akaaiarvi* (November), "The Eskimo lay down food depôts."
13. *Hikkernillun* (December), "The sun disappears."

The difference between this and our own division of the year into months is not so very great.

The seasons are computed according to the conditions of the ice and snow

Opingan, spring (June and July), the season when the snow disappears from the ice, and the ice breaks up.

Avra, summer, the season when there is no ice, August and September.

Okeo, winter, the remaining eight months.

They have thus only three seasons. They do not recognise any autumn.

The twenty-four hours of the day are divided into :—

Obla, morning.

Onon, evening.

Onoa, night.

As an instance of the accuracy with which they keep count of time, I may mention that, on March 25th, 1905, Talurnakto told me it was now a year since we first met. As a matter of fact our first meeting with the Nechilli took place on March 18th, 1904. This was not such a bad computation, without a diary or almanack.

As regards the religious ideas of the Eskimo I will not venture to give any account. The statements I could obtain on this subject were exceedingly imperfect and vague, and left most things to one's own imagination. If these people had any belief in a higher being they at any rate concealed it very jealously. They imagine a life after death; at any rate good men are assigned an abode in the moon, the bad in the earth; the stars are destined for those who had something of both in their nature. Natural phenomena, such as the aurora borealis, shooting stars, thunder and lightning, rainbow, etc., they regarded with complete indifference. Evidently they loved life, but on the other hand they had not the slightest fear of death. If they were sick or in misery they bade farewell to life with a tranquil mind and strangled themselves. Two such cases occurred during our sojourn among them. During the voyage of the "Gjöa" we came into contact with ten different Eskimo tribes in all, and we had good opportunities of observing the influence of civilisation on them, as we were able to compare those Eskimo who had come into contact with civilisation with those who had not. And I must state it as my firm conviction that the latter, the Eskimo living absolutely isolated from civilisation of any kind, are undoubtedly the happiest, healthiest, most honourable and most contented among them. It must, therefore, be the bounden duty of civilised nations who come into contact with the

NECHILLI ESKIMO VISITORS ON BOARD.

Eskimo, to safeguard them against contaminating influences, and by laws and stringent regulations protect them against the many perils and evils of so-called civilisation. Unless this is done, they will inevitably be ruined. All honour is due to the Royal Danish Trading Company for the manner in which it has treated its Greenland Colony. It is to be hoped that other nations will follow the Danish example in this respect, and will be fully alive to their responsibility in regard to these splendid and doughty children of Nature up at the Pole.

My sincerest wish for our friends the Nechilli Eskimo is, that civilisation may *never* reach them.

CHAPTER IX.

FAREWELL TO GJÖAHAVN.

WE noticed, with the most lively interest, that as the spring days were approaching, the uniform snow, which surrounded us on all sides, was little by little becoming streaked and marked with the tracks of animals of various kinds. We had seen a few tracks of foxes all through the winter. One morning the first ptarmigan had been there, executing some of their weird and perplexing joy dances, with their tripping downy feet. Another day there was a fresh pattern among the others, the track of a lemming, the first we had seen; it was straight at first, then zigzag, suggesting fear of an enemy, then mixed up confusedly with fox tracks, and here and there small red blood stains on the white snow. We watched the changing surface of the snow like a chart of approaching spring. On board we had the snow and ice removed, and all the skylights and ventilators opened. Light and air flowed in and freshened the "Gjöa's" stuffy winter room.

We had now made such progress with our Eskimo friends that we could talk and chat with them on any subject. It must not be imagined that we could speak the Eskimo language by any means; we were probably further from speaking it after living with them for two years than we were in the beginning. If, for instance, we met any strange Eskimo, as Ristvedt and I did in the sledge trip in 1904, or as the Lieutenant and Ristvedt did on Victoria Land in 1905, they could

Kimaller. Lindström. Anguju. Errera.

LINDSTRÖM BEING INSTRUCTED IN THE NECHILLI ESKIMO METHOD OF BUILDING SNOW HUTS.

not understand a word of what we said, or we of what they said. But with our old friends, as I have said, we had formed a sort of language of our own, by which we could easily understand one another. I say "we," though I only mean six of us, as the seventh obstinately persisted in his contempt for the foreign tongue, and always used his own, but the Eskimo understood him all the same. They were at times more intelligent than many white men.

Talurnakto was by this time quite one of ourselves, and, in order to have him within reach at any time of the night, we let him lie on the floor in front of the Lieutenant's bunk. He snored so loudly that he woke up the whole ship's crew, so the Lieutenant took him in hand, and, at the first sign of snoring, he let fly a boot at his head, whereupon he turned over and grunted. Then he would begin again and get a fresh reminder; at the third or fourth reminder, as a rule, he stopped snoring. When the Lieutenant left the ship, Talurnakto, much to his satisfaction, took the vacant berth. As he lay in the bed, his round face outside the blankets, with his evening pipe lit, and smoking rank chewing tobacco, Talurnakto looked the picture of content. The ventilation was so good, that the quality of his tobacco did not trouble me, but when he laid down his pipe and turned over to go to sleep he was soon snoring so loudly one would have thought his head would burst. I had previously provided myself with missiles, and the struggle began; books, boots, and the like flew across the cabin, until at last it ended in Talurnakto putting his head out and saying "Go natti" (good night); he then settled down and slept quietly. Talurnakto knew more Norwegian words than our aforesaid comrade knew of Eskimo.

On going through the magnetic observations made

round the stations a doubt occurred to me as to the possibility of our observatories being situated too near to the ship, and that the large quantity of iron on board might have exerted some disturbing influence. Wiik and I then made numerous observations, the result of which removed all doubt on this point. The distance between the ship and the nearest observatory was about 500 yards, and this proved to be quite sufficient.

There was a good deal of other work in store for us this spring; we had to get ready to leave Gjöahavn, and our houses had to be taken down as the boxes used in building them would have to be again utilised for packing the tin cases in. All the instruments had to be packed up, the stores brought on board, and the ship itself made clear. We should have to wait, however, until the spring really set in, and this would not be before the month of June. In the meantime fresh crowds of Eskimo were arriving, and among them were a good many strange faces, attracted here by the report of the great treasures to be found in Ogchoktu. Many of them had travelled several hundred miles to reach here. They had very little with them as they could not travel with heavy loads on these roads, but we gave them wood and iron in exchange for the few seal-skins they offered us, and they left highly contented. Umiktuallu was a keen business man. He had noticed that I liked to have neatly sewn clothes, so he bought garments from some of his friends and sold them to me at a large profit. In the course of the winter he had procured some powder and shot to use with his muzzle loader. Since then, however, his brother, the "Owl," had lent him the Remington rifle I gave him, and now Umiktuallu wanted to change his muzzle-loading ammunition for Remington cartridges, so he came to me and asked if I would change it. He was a very clever hunter, and as he promised me the meat of the deer he killed I accepted the proposal. A couple of days after he came on board to settle the business, and with the most innocent look in the world handed me over all his shot but kept back the powder. I pretended not to notice anything and went to fetch the Remington cartridges, leaving Umiktuallu grinning all over his face with glee at having, as he thought, so successfully tricked me. I counted out the 400 cartridges, and then quietly began to take out the powder. He stopped me and said that he had no use for the balls without the powder and caps. "No! No! No!" said I, "that is just my case with you and your shot." He then produced the powder and the exchange was effected, but Umiktuallu could not help laughing at his own failure.

One day the inventive Lund surprised me by showing me a new gun. It was his masterpiece, as an inventor, and was by no means badly constructed. Of course, Ristvedt had had to give him a hand, but this did not detract from the inventor's merit; the weapon was really a curiosity and might have claimed a place in any museum. The barrel was a piece of iron piping which belonged to the petroleum tanks in the engine room, and the gun lock was an old one he had found in the old depôt at Beechey Island and had taken care of. Strange to say on this lock was the name of Andreson, Gunmaker, Tromsö. It is impossible to say how this article got there, it may have been brought by one of the Scotch whalers that occasionally put into Beechey Island. Lund had now an opportunity to utilise it. He made the stock of the gun out of birch wood, of which we had a few pieces on board. I cannot say that the weapon was exactly a handsome one. It was a muzzle-loader, and we were all very excited to see how the trial would pass off. The gun was placed some distance out on the ice and a string tied to the trigger, and from thence taken to the ship so that it could be fired off from on board. The trial turned out quite successful, and Lund had the satisfaction of presenting to his friend, Uchyuneiu, a gun of his own manufacture. Uchyuneiu shot several deer with this weapon, but said that he had to get very near to kill them. Probably the bore was not quite true.

One day the "Owl" came on board as happy and contented as ever, and had a great deal to tell us about the seal fishing. The Eskimo were now catching as many as sixteen seals a day, and passed every night dancing, playing, and eating as much seal flesh as they could stow away. I invited the "Owl" down into the cabin and there we sat, he, Talurnakto, and myself, chatting like good old friends. Suddenly, Talurnakto told me that the "Owl" had broken his collar-bone;

I examined him and found this was correct; the upper part of the collar-bone having been fractured by a fall on the ice. With Wiik's help I bound up the fracture and enjoined the "Owl" to keep quiet for fourteen days, and in order to control him I offered to let him remain on board during that time. He gladly accepted. Talurnakto was delighted to have the society of his good friend, and these two passed many an hour together over the "Fortress" game.

Easter was now approaching, and the cook made great preparations for it. On Maunday Thursday my old friend Atikleura came with his wife and child. We had not seen each other for a year. He brought with him the gun I had given him, polished, cleaned, and brightened like new, and very carefully kept in a deer-skin case which his wife had made for him. About the time of the breaking-up of the ice in the previous year, he had gone over to Nechilli with his old father, Kagoptinner, and family, to pass the summer and the first part of the winter there. I ordered salmon from him then, and he now came with seventy large fat salmon caught during the autumn at Nechilli. He also presented me with two large parcels of seal blubber, four deer skins, and a bear skin. In exchange for this he received a saw, a hatchet, a knife, and a hundred cartridges. He beamed with delight. Nalungia received needles, a thimble, some beads, and matches. Talking about beads, Lieutenant Hansen and I had brought with us a large supply of beads in the hope of being able to do some business with them. Our disappointment was therefore very great when we saw that the Eskimo women did not care a fig for this style of adornment. But Lieutenant Hansen's business talent came to his assistance here. This was the first and I believe the only time I ever perceived that he had this talent. He began to make rings, brooches, necklaces, and other things out of the beads, and one day we had a feast in the cabin, and invited the Eskimo to partake of golden syrup and dog's suet. When the merriment was at its height the Lieutenant distributed some of the ornaments as presents to his guests, at the same time showing them some photographs of Greenland Eskimo who were covered with these trinkets. From that moment beads went up in value, and when we left Ogchoktu, their market value ruled very high.

One day, when I went over to visit the Eskimo, I met in their camp a young Ogluli boy, named Ugvi, who had taken part in the robbery of the store tent in the winter; as I had strictly forbidden the thieves to come to Ogchoktu, he had till then kept himself concealed in the huts, but now, as bad luck would have it, he ran right into me. I took hold of his arm and said severely, "Is that you, you little thief, don't you know that I have forbidden you to come here?" The boy did not move a step or change a feature. He simply stood and looked at me with a sly smile, but some of the older Eskimo came and assured me that the boy should be sent away immediately. Then I went to Atikleura's tent, and shortly after he himself came in with the boy, who happened to be a brother of Nalungia. She was extremely distressed on hearing he would have to go away, and it was very touching to see how carefully she packed his provision sack for him with salmon and other good things. I now saw that our authority in the locality was absolutely undisputed, and in consideration of the fact that we had not very long to stay there, I informed Atikleura that the young rascal might remain if he, Atikleura, would be responsible for his good behaviour. This he immediately agreed to, and everyone was satisfied. As a matter of fact, my own indignation at the theft was not very great; it was done under stress of want and hunger, and would not have happened under ordinary circumstances. However, I enforced my prohibition as regarded the adult thieves, and so, when old Teraiu and his family came to settle amongst his friends, he heard that the prohibition had not been removed, and he immediately left the place. I mention this to show with what respect we had succeeded in inspiring these people, and, without boasting, I may say that the result was obtained without brutality or harshness, but simply by always upholding justice and right.

Poor little Kabloka, the "Owl's" excellent little wife, who was always ready to do whatever she was asked, was now, after three years of wedded life, in an interesting condition. Notwithstanding this, she had travelled twenty miles a day with her husband. Presumably proud of the prospect, she had adorned the most prominent part of her little body with a number of beads, and her appearance was consequently very comical.

It was only now that we were on such good terms with the Eskimo that they really trusted us and imparted their confidences to us. I had often asked them if they knew anything about Franklin's Expedition, but I merely got an evasive answer. At length, however, Uchyuneiu, the Ogluli Eskimo, told me what he knew. He was a very brave and intelligent fellow, and his account agreed very well with that which Schwatka had obtained twenty-five years ago. One of the ships had driven down towards Ogluli and was found by the Eskimo one winter's day when they were seal fishing on the south coast of Cape Crozier, the most westerly point of King William Land. They had then removed all the iron and wood work they could remove, and when spring came and the ice broke up the ship sank. At that time the Eskimo had eaten something from some tins which were like ours, and it had made them very ill: indeed some had actually died. They knew nothing of the other vessel; in all probability it had been crushed by the ice on the north side of the Royal Geographical Society Islands. In accordance with this information we could almost safely say that the only unnavigated portion of the north-west passage, extended from the point where this vessel sank to Cambridge Bay on Victoria Land, where Collinson wintered in 1852.

One day Talurnakto came on board radiantly proud and told us he had given Atikleura "a black eye." This astonished and at the same time interested me, so I endeavoured to ascertain how it had happened. Yes, it was quite correct; a crowd of men were gassing outside Atikleura's hut when Talurnakto made some remark that offended Atikleura, who then boxed the fat little fellow's ears. Under ordinary circumstances Talurnakto would have taken his punishment quietly, but now he was half a white man; he lived on board, and was altogether a man of honour and dignity, so he could not afford to take such an affront, and consequently he struck out at Atikleura with all his might, hitting him in the eye. There the matter ended, blow for blow, and that was enough. The Eskimo regard these things with the greatest coolness.

We were now getting towards the end of April and the sun had already melted the snow at various places inland. As in the previous year, the sight of the bare fields was an indescribable enjoyment to our eyes and to our feelings. On May 1st the ice measured six feet in depth, as compared with twelve feet six inches in the previous year. From April 1st to May 1st this year it had decreased nearly an inch, whereas during the same period in the previous year it had considerably increased.

There were very few ptarmigan this year, indeed it was all we could do to obtain a sufficient supply for one meal a-week. On the other hand there were plenty of foxes. Ristvedt caught them in traps and probably would have caught many more if he had had the time to attend to his traps more frequently, but this he could not do, and it often happened that he found that the remains of the trapped animal had been eaten up by its own relatives. Now, in the spring season, the foxes had a peculiar habit of going out on the ice to find out seal holes ; here they rummaged about as well as they could in the snow but did not get much more than a smell of the seal.

When leaving home we had not taken with us any shot guns, as we had been informed that we could easily get them in Godhavn. Ristvedt was the only one who had brought his fowling piece with him. When we arrived at Godhavn we could not get a gun for love or money, but the Governor and his assistant were kind enough to lend us theirs. I obtained a single barrelled gun, and Lieutenant Hansen a double-barrelled gun, although we were told they were not very good weapons, but something was better than nothing, and we got some use out of them. The Lieutenant had a good deal of trouble with his gun ; he often shot the barrels off, and eventually it looked more like a "shot revolver," and was christened the "garden syringe." In the autumn of 1904 he put the gun away altogether ; then one day Hansen found it, cleaned it up, and gave it to the delighted Talurnakto, who felt sure he would make good use of it, if he could only get near enough to the birds. One afternoon Talurnakto came on board in tears. The tears ran down his face as he told us that the "garden syringe" had burst. He had fired off both barrels at once, and this was more than the old piece of furniture could stand. When he saw that we took his misfortune very lightly, inasmuch as we told him that he was lucky to get off with a whole skin, his tears very soon changed to smiles.

My credit with the Eskimo was really very flattering. We were quite out of knives, our best bartering medium, although at the time we had four large excellent ice saws of steel, and a great number of knives could be made from these. As, however, our cutler was out on a sledge trip, I had to give out warrants for knives for future delivery, that is fourteen days after the return of the cutler. Also, as regards ammunition, I was in the habit of giving delivery warrants, payable by the smith, who also had charge of the ammunition. In the beginning the Eskimo were rather astonished at receiving a piece of paper instead of a knife or fifty cartridges, but when they understood the meaning of it my paper was always accepted as good payment. I made out several of these warrants in the summer of 1904, and some of them were only presented in the summer of 1905, but they were immediately honoured to the great delight of the possessors.

In the beginning of May all the women suddenly seemed to have got the peculiar idea that I set an exceptionally high value on seal bladders. They used these bladders for preserving reindeer fat during the summer. Now they came in great numbers to the ship with whole heaps of these bladders blown out. For a time I accepted them, and gave them a few needles in exchange, but at last it became too much for any one man, so I had to tell them that there was no further market for bladders. It was obvious that one of my comrades must have put them up to this, and the result was that afterwards the whole cabin was covered with these blown-out bladders.

On May 9th we seriously began our spring work. In this we were considerably helped by Talurnakto, who, by the way, had just appeared in a new costume. Wiik had given him a pair of sealskin trousers from Godhavn, and they fitted him like a skin on a sausage, so that it was all he could do to get them on in the morning ; however, he wriggled in, then he bent his knees slightly by way of experiment (bending them much was out of the question), and he declared with satisfaction that they would make splendid kayak trousers. In this he was correct, as there would be no difficulty in keeping still in such garments.

Half of the sail roof over the vessel was removed, and everything put in order to clear up the hold. Lindström's underground sojourn was now at an end. The galley, which had been all along in the hold, was now again placed on deck. It was very nice to see the old box in its place again, although its looks had by no means been improved by its sojourn down in the dark hold. In the course of the winter a good deal of damp had collected in the hold, and we pulled down the draught screens to dry it. Fifty casks of petroleum which were lying in the bottom of the boat were now emptied into the fixed iron tank. The empty casks were brought on shore, and there they stood, the admiration and ambition of the Eskimo. It was my intention to make all our empty casks, boxes, tins, old boards, and the like into ten heaps, and to give a heap to each of the ten best and most capable Eskimo; but for the present they had to content themselves with the sight and the odour thereof.

We first brought on board ten tons of large stones for ballast. Now, in order to bring the stores on board, we had first to procure cases, consequently we had to dismantle our houses to re-acquire the materials used in erecting them. We could dispense with the "Magnet" most easily, but I wanted to retain the "Variation House" as long as possible, so as keep the observations going. Wiik now moved on board with all his traps and again took up his old place in the cabin, and on May 15th we commenced pulling down the "Magnet." The whole station, with all its little houses, had become as dear to us as a home, and it was with much regret that we dismantled it. As the Magnet Hill lay there bare and desolate as it was when we came, it looked more solitary than ever, but behind this sad feeling there was another stronger and brighter; it was the beginning of our further progress. Every case we carried down brought us nearer to the breaking up and nearer to the goal of all our hopes and longings. I cannot deny that I burned at the thought of the time when we should show our Norwegian flag to the first vessel on the other side of the North West Passage. All sadness and sorrow yielded to this desire, and we pulled the cases down with impatient eagerness.

SCENE ON DECK (SUMMER, 1904).

Farewell to Gjöahavn.

Saturday, May 20th, was a great day on board the "Gjöa." Till late in the evening the day passed as usual. At 12 o'clock noon the sun stood in the south as usual; at 3.30 in the afternoon Lindström had finished his usual siesta; at 6.30 we took our supper and intended retiring according to custom at 9.30. Wiik came from Magnet Hill, where he had been taking meteorological observations, and told us that some people were coming along on the ice; this was not an unusual occurrence, but as he thought he had noticed, in spite of the great distance, that it was a sledge with a lot of dogs, and that it was travelling very rapidly, I sent out Talurnakto to see what it was. Our Eskimo friends do not, as a rule, travel "express," and as Talurnakto did not return, we felt sure that it must have been an Eskimo family now stopping to camp for the night. I therefore got into my bunk, but I had not been there very long before I heard hasty and unfamiliar footsteps on deck, and immediately afterwards a man burst into the cabin. "Go' morning! You give me 'moke!" was his greeting. It was Atangala, with his broadest and most triumphal smile. He stood before me and held out his hand, at the same time asking me not to press it too hard as he had hurt it. I cared little for his smile, hand, or 'moke, I merely wanted to see whether he had the mail. "Have you any letters?" "Letters? Yes, out there on the sledge, a whole heap of them." He was both surprised and hurt that we were in such a hurry about these letters, but I hastily put on a few clothes, and we both went outside. In a hurry and scurry everyone got up, and soon we stood beside Atangala's sledge ; at last he brought out a neat little soldered tin box from beneath all kinds of odds and ends.

This was the mail!

I shall not endeavour to describe my feelings when holding this tin box in my hands, containing as it did messages from the living tumultuous world. We well knew that there could not be any direct message from the dear ones at home, but here, at least, was news of the great human community to which we all belonged, and from which we had so long been cut off. The simple word "mail" produced an indescribable sensation in us all. We carried our treasure on board and gathered around it. Lund immediately got the soldering-lamp to work, and the box was soon opened. The first I found was a letter from Major Moodie, Chief of the Royal North West Mounted Police, and Chief Commander of "The Arctic," belonging to the Canadian Government. This ship was formerly the "Gauss," which had been built and used for the German South Polar Expedition, under the command of Erik von Drygalski. "The Arctic" was investigating the conditions around Hudson Bay, and had wintered off Cape Fullerton, near Rowe's Welcome, an arm of Hudson Bay. In this very friendly letter he offered us every assistance, should we go near him, and he also sent me five sledge dogs. From the Captain of "The Arctic," Captain Bernier, I also received a long and interesting letter. His information about the

Godfred Hansen. Gustav Wiik. Anton Lund.

FESTIVITIES IN THE CABIN.

Roald Amundsen. Helmer Hansen. Peder Ristvedt.

FESTIVITIES IN THE CABIN.

American whalers on the north-west coast was very acceptable to me, and of the greatest importance. The Captain also sent us a quantity of newspaper cuttings and photographs, which we greedily devoured. The American whaler, "Era," was wintering at the same place, and the commander, Captain Comer, sent me a warm and friendly letter. He also sent me five dogs. Hoping that Atangala might arrive before Lieutenant Hansen's departure, I had asked for dogs, but as they arrived now six weeks after the departure of the expedition, and as we had no food for them, I was compelled to send them back. Major Moodie very kindly undertook to forward our mail home. I am unable to find words to thank these three gentlemen for the kind helping hand held out in such a warm and friendly manner.

Of all the news we obtained, the most interesting was that about the war between Russia and Japan. We were glad and thankful to learn that the Danish Government had established a depôt for us in Leopoldhavn, which might still be of some use to us.

We sat up late through the night and discussed all this news. In our hurry to get up we had not troubled to dress very much, and, as we sat in eager groups around the letters and newspapers, we formed a very amusing picture. Atangala took advantage of the time and opportunity, and had one "'moke" after the other. This time he was accompanied by his son, Arnana, a young man about twenty-five years old, a very fine fellow. When at last I had time to attend to Mr. Atangala, he told me he had had a very fine trip, although he had run rather short of food until he got near his home at Chesterfield Inlet. Of course, he had to go home first. Here he had encountered some musk oxen, and had hunted them; one day, when out shooting, one of the cartridges had jammed in the barrel of the gun, and, in endeavouring to get it out, the cartridge exploded, with the result that Atangala lost his fore-finger. When he reached home his friends and relations tried to persuade him to give up carrying the mails and to remain at home quietly and take care of his finger; but he withstood all temptations and continued his route. The recompense he had in view, after successfully accomplishing it, was an old gun and four hundred cartridges; but I am inclined to believe that it was not merely the payment which impelled him to fulfil his task, but that he was stimulated by a desire to prove that he was a man of his word, and in circumstances like ours, such a man is doubly appreciated. He consequently received a considerable addition to his pay, and during his entire stay with us he was treated as the honoured guest. He was especially delighted at the praise I gave him for his integrity and sense of honour. On May 23rd at 11 o'clock in the morning he started out again for the south, together with the superfluous dogs we were sending back, as also with the return mail. He was to try and reach "The Arctic" before she could sail, that is, before the ice broke up.

In the meantime we were at work with the stores. The tin cases were removed and placed in wooden boxes, numbered and entered in the store list. In two days this was all ready and for the second time—presumably the last—our stores were ready to take on board. The Eskimo were very useful to us, and there were generally a number of young fellows doing nothing whom we immediately engaged. With the aid of three Eskimo I saw to the work on land, whilst Lund and Hansen directed everything on board. The work progressed rapidly, the cases were carted out on sledges to the ship's side, hoisted on board, lowered into the hold and then stowed away. An exact plan was made of the stowage so that we could find whatever we wanted at any moment.

The deer had begun to return, and Talurnakto was constantly hunting them. In the afternoon he returned home and then went off with the sledge to bring in what he had shot. I had bought a dog from the Nechilli Eskimo; its name was "Achkeamullea"; a very easy name to call out in a hurry, isn't it? It occurred to me it was rather long, so I rechristened him "Akchya," Northern Light. I also kept one of the dogs Atangala brought; his name was worse, and I rechristened it "Fix." Fix was as thin as a skeleton when he came, but in a fortnight he was so fat he could scarcely move. We fed both the dogs on the reindeer offal and used them for bringing in the deer shot by Talurnakto. However, during a fight with Fix, Akchya was ripped open, and died soon after. It was a shortsighted act on the part of Fix, as he now had to do all the work himself.

Chapter IX.

At the end of May the Eskimo concluded their winter seal fishing, and returned to land to their summer occupations, hunting and fishing. On their way they came in large crowds to Ogchoktu, to visit us and, if possible, to do some business with us. I utilised the occasion for distributing our recompenses among them. My ten heaps of wood and iron had been increased to eleven, as there were, in fact, many who had been very loyal to us all along, and had helped us assiduously and industriously, and who might be assisted with advantage. It was a great pleasure to see how happy they were with their gifts; Talurnakto, especially, was quite radiant; he walked up and down past his treasures with the dignified and grave air of a rich man. Poor fellow! If he again goes into Akkla's hut as co-husband all his riches will soon disappear.

On June 1st, 1905, the self-registering instruments were stopped, after nineteen months' uninterrupted working. Wiik had carried out this work all by himself, and did it with a care and accuracy beyond all praise. This was Ascension Day, so no work was done. In the afternoon I collected together all our Eskimo lady friends to enrich them with our empty tins. There were some hundreds of tins, and I had put them together in a large heap in the middle of the hill. Then I had the womenfolk arranged round the heap in a ring and told them that when I had counted three, they might "go for" the heap and get all they could. The men arranged themselves behind their ladies: One! two! three! and in they rushed, using both hands and shovels; they threw the tins out backwards between their legs—they were not hampered by skirts—and the men grabbed hold of the flying tins, and so each collected his lot. Laughter and noise, shrieks and shouts, tins flying, men rushing, and so the heap was cleared.

The same afternoon there arrived a family consisting of husband, wife, and three sons. The eldest of these latter was a young fellow, twenty-five years old, who had been lame from his childhood, and the parents drew him along on a seal skin, as they had done for many years. He was a bright fellow, slender, and well proportioned, with long jet black hair. He was tattooed over the nose, the only tattooed man I saw, and was looked up to as a magician. As we had more sledges than we needed, I presented these people with one of them, so that it would be easier for them to move the invalid about. This made them indescribably happy, and the young man himself hardly knew how to express his gratitude. He then gave me what he valued most, his magician's sign, a crown and a brow-band of deerskin, which I highly valued, as it made a very notable addition to my ethnographical collection.

Navya was re-engaged as needlewoman on board; she came after breakfast and remained till evening, and had her meals on board. Her little son, Nanurlo, came with her. She sat in the cabin and Wiik had her company when he was engaged entering his magnetic observations in the journals. In spite of her fifty years,

Chapter IX.

Navya was always as bright and cheerful as a young girl, full of nonsense and an awful chatterbox. One day when he had been called up on deck, and came down again after a few minutes business, Navya had "taken up the pen" and scribbled all over the page of the journal, spoiling all his neat rows of figures. She was very proud of what she had done, and was surprised by the flood of invective Wiik let loose on his return; but it was impossible to be really angry with Navya for long, and Wiik soon forgave her. She was exceptionally clever as a needlewoman, and made clothes of all my fine skins; of the bear skins she made trousers, and of the fine reindeer skins she made coats. For every article she made she received some small object as payment. What she preferred, indeed, what all the ladies seemed to prize most, was our enamelled ironware, as well as china and earthenware. They were very fond of the white, and they would sit in the huts for hours licking and polishing these articles.

The small boys had now, as in the previous year, begun to take their places again on the ice, where they fished for codlings. Now and then I went among them and could not restrain myself at times from asking them to lend me the line to try my luck. This seemed to amuse them immensely, as they could not imagine a grown-up man fishing for codlings of his own free will.

On June 2nd we began to pull down the "Variation House." Lund and Hansen were entrusted with this work, and with the assistance of the Eskimo they made rapid progress.

When I was leaving the "Deutsche Seewarte," Professor Neumayer had given me a photograph, which I promised to place as near the Magnetic Pole as possible. On my sledge trip to the Pole in the spring of 1904 I had the photograph with me, but was unable to find a convenient spot to place it on. The Eskimo would, at the first opportunity, have plundered everything we buried, if we had indicated its position by any sign. I therefore took it with me to bury it on the spot where we had carried out our chief work, beneath the "Variation House." As the house had now been removed, and the Eskimo had ransacked the site, I took advantage of a moment when no one was there and buried it. On the back of the photograph I wrote: "In deep gratitude and respectful remembrance I deposit this photograph on Neumayer Peninsula. "Gjöa" Expedition, August 7th, 1905.—Roald Amundsen." This was placed in a flat tin case and buried under the cemented stone foundation on which the registering instrument had stood, and was then covered in with sand. An oblong heap of earth of the same shape as the house marks the spot where the "Göja" Expedition had its self-registering instruments for nineteen months. If any future expedition should come here and desire to establish its instruments on exactly the same spot, this tin case may serve them as a guide. Later, the situation of the house was accurately determined by measurements, by Lieutenant Hansen. On the site where our observatory for absolute magnetic measurements had stood, just under the centre of the instrument stand (which had occupied the same position the whole time), we buried a piece of yellow limestone about one foot long by six inches wide, with the letter G cut into its upper surface. With a little searching it will not be hard to find it again. If we had marked the spot by a cairn, the Eskimo would probably have respected it for some years, but there is no doubt they would have overthrown it eventually. Now that everything was on board, we had to smarten up the ship, as we might possibly be meeting people this year, and I should not like them to say that we Norwegians do not keep our ships in good trim. The "Gjöa" and the boats were therefore painted and oiled all over, and they looked as well as they did the day they left the wharf. All shared eagerly in this work, and Ristvedt and Lund also made an elegant accommodation ladder of iron and wood, which attracted great attention when we arrived at San Francisco. At any rate, no one shall be able to say, from the state of our vessel, that we have wintered for two years.

One June day I went to the Eskimo encampment and saw one of them sitting and eating with one of our table knives. I immediately took it from him and asked him where he got it. He told me it was a present from Talurnakto. I immediately summoned the culprit before me, and he admitted that he had taken the knife. It had been accidentally thrown overboard, and he had appropriated it. He did not endeavour to excuse himself, and it was not a very serious matter, yet for the sake of example, I forbade him to come to the ship for eight days. He went off and strictly obeyed my injunctions. When the time was up he came back as smiling and contented as ever.

We were now eagerly expecting the return of the sledge expedition. As they had not returned after the lapse of the first seven weeks, I concluded they had found the depôt on Cape Crozier in order, but, even so, they should have been back by June 16th. Many Eskimo had just come all the way from Ogluli, but none of them had seen anything of the Expedition, and, consequently, I began to be a little anxious. Since the arrival of the Eskimo in such large numbers we had begun to have a night-watch. I instructed the watch to keep a sharp look-out, and immediately to inform me if he saw anything which might indicate the arrival of our travellers.

The 24th, St. John the Baptist's Day, was to be a holiday, and, as we had fixed Thanksgiving Day for the same date, it became a holiday in a double sense. At 6.30 in the morning Lund, who had the watch, came and woke me up, saying "Here are the boys." I was not long dressing. It was a splendid morning, perfectly calm, with a burning hot sun, and there were our two comrades approaching from the direction of Fram Point. I can hardly say how glad and relieved I was to see them, and the rate at which they drove the dogs went a long way to show that the animals were still in good condition. The flag was immediately hoisted, together with all the bunting we had on board, and everyone was very festive. A more elaborate breakfast than usual was prepared, and as we ate it we soon got a brief outline of all the incidents of the journey. The Expedition had furnished them with hard as well as lighter work, but more of the former. The depôt at Cape Crozier had been wholly destroyed by the bears, but they had succeeded in killing four reindeer. The passage across Victoria Strait had been exceptionally difficult. The ice had been trying and the surface very irregular; some days, indeed, they did not cover more than two or three miles, and they repeatedly had to go out of their way. On the other side of the Strait they had encountered a new tribe of Eskimo, the Kilnermium, from Coppermine River; they were out seal-fishing. Whilst these Eskimo were almost destitute of iron they were better provided with wood than the Nechilli; their boats and sledges especially were better. For nails and small knives they bartered as much seal flesh and blubber as our comrades needed. They passed one day with these people to get a rest, which both man and beast had need of after their arduous work on the ice. They then continued their way along the unknown coast of Victoria Land. The land was so low and flat that for the most part it was difficult to distinguish the coast from the ice. They charted the coast as they went along. They were constantly shooting seal, reindeer, and bear, so that they had ample stores practically all the time.

On Friday, the 26th, they turned back after having constructed a cairn at the most northerly point they

REJOICINGS.

reached, and deposited a report in it. The return was easier and more rapid, as they had no surveying to do. On the return journey, the land seen by Dr. Rae in Victoria Strait was carefully investigated. It proved to be one group out of a mass of small islands—"The Royal Geographical Society Islands." These islands were charted as well as possible, and later we found this to be of great importance to us in our further navigations. The Ogluli Sea, between America, Victoria Land, and King William Land, turned out to be very full of islands and not clear and free as shown in the old maps. It is important to bear this in mind, in the event of navigating that part on a dark night.

On the return journey to Cape Crozier they were lucky enough to have better ice, so that they made more rapid progress. With the exception of a few sore feet among the dogs, everything was in the best of order. The trip had occupied eighty-four days, although they had taken with them stores for only fifty days. The result was excellent, I might even call it splendid, considering the many days of bad weather they had, all the careful surveys they had made, and also the fact that they had had to devote a good deal of the time to hunting to keep up supplies. We learned all this during the first breakfast; after that, the day was spent in rejoicing and festivity.

Towards the end of June it got very warm and the channels began to open up. If it continued like this we might have as favourable an ice year as in 1903. The land was almost clear of snow already and the gnats were tormenting us considerably.

On board we had to make various changes in our arrangements after Ristvedt and Wiik had moved over from the land. The Lieutenant and myself had to share the cabin with them. As Wiik was now continuously occupied in carrying out his observations, the cabin could no longer be used as a dark room; but as it was absolutely necessary for the Lieutenant to have a dark room, the problem was eagerly discussed and many proposals were made for its solution. Lieutenant Hansen finished it at length by using as a dark room one not originally intended for that purpose—I need not particularise further—"necessity is the mother of invention." The store tent was now empty, as all the stores were on board. This tent was now used for many different purposes; first as drying room for all our bird-skins. These were hung up in it and soon dried in the continual draught. Afterwards it was used as office and bath-room, and the price of admission was so modest that it did not prevent anyone from washing himself. All that was required was to light the "Primus," see that it was put out afterwards, and leave everything in order. There was no bath attendant. We also arranged some long boards on which we set out our observation books and magnetic curves in order to go through them for the last time, and also to air them before they were finally soldered down.

The "Owl," Umiktuallu, and Nulieiu, started on June 2nd with their families westward to Kamiglu near Eta Island in Simpson Strait, intending to go deer hunting. I arranged with them that we were to stop the ship when we were passing there to take on board the reindeer meat. We gave them a long pole with a flag to fix up so that it would be easier for us to see where they were.

I had long promised Lindström he should have a holiday one of these days to investigate the land in the mystic interior. The Eskimo had talked much about the river up towards the north, Kaa-aaga angi (the large river) which, according to all reports, was swarming with salmon. Several families had gone there, and had reported that the fishing was splendid. So Lindström fixed on Kaa-aaga angi, and his longing to leave his galley for this river got stronger and stronger. At length on July 4th he was ready. The expedition consisted of himself and Talurnakto as second in command. For a long time the others had referred to it as the "Expedition for Investigating the Interior of King William Land." I did my best to preserve a serious demeanour with regard to this expedition. Lindström's real object was to increase his zoological collection, and his comrades made great fun of him, but no one could help laughing at these two Arctic explorers as they moved slowly across the ice under a shower of more or less complimentary observations from the boat. They were both of similar build, tallest when lying down, and they trundled off side by side, like two balls. Next day, Ristvedt and Wiik, who when they got together, were capable of inventing the most incredible tricks, went up to the Eskimo Encampment, and selected a walking specimen of the same type, a youth of the name of Tonnich. This plump little fellow was provided with a supply of stores containing bad tinned meat, and was told to keep on till he came up with Lindström and Talurnakto. He took with him a long solemn letter to Lindström expressing deep anxiety as to the "great scientific expedition" he was leading,

LIEUTENANT HANSEN AS PHOTOGRAPHER.

and at the same time hoping that the relief expedition under Tonnich, the Eskimo, might find the two brave travellers alive, and rescue them before it was too late. When they returned on July 9th, they were received with great ovation, and all kinds of undesirable attentions. The result of the expedition consisted of forty eggs and a few eider ducks. The scientific report was short. Kaa-aagȧ angi was a river about as broad as the Nid River, at Trondhjem, and the fauna and flora were richest at the station. Lieutenant Hansen had in the meantime been over to the Duke of Abruzzi's Point and determined its position, but it is not pleasant to go for sledge trips at this time of the year, and the Lieutenant returned soaking wet.

In the the middle of July I discarded Eskimo apparel and dressed again in my own clothes. I had worn Eskimo dress continuously for twenty months, and had had an opportunity of forming an opinion as to its utility. For winter use in these regions, as previously stated, light deerskin clothes are the best, and they ought to be made loose, although during the most intense cold and towards the springs, like the Eskimo, I wore a broad belt of reindeer or wolf skin around the abdomen, to keep the cold air from this sensitive part. In summer, on the other hand, it is not advisable to wear deerskin clothes, as they get soaked through when it rains. I came to the conclusion that woollen underclothing with sealskin outer clothing was the most suitable for summer wear. I always wore deerskin stockings with the hair inwards. I put sedge grass in the feet of the stockings, and outside these I wore deerskin boots with the hairy side outwards. Thus protected, the feet cannot be injured in any way. The sedge grass absorbs the moisture, so that if the grass is taken out at night and dried, the boots will be kept pretty free from damp. To make quite sure, the stocking may be turned inside out at night.

At the end of July the Ogchoktu Swimming School was started; Wiik and Ristvedt being the teachers, and Hansen and Lund the pupils. A fresh water pond near the boat was used as the swimming-bath, and here they bathed every evening as long as the weather was warm enough. The temperature of the water was about 59° Fahr., which was not bad for Polar regions. In the beginning they used swimming ropes, but, as the pupils were rather quick at learning, they were soon able to dispense with these. Sometimes the Eskimo came and looked at them, and went away convinced that the Kablunas were mad. One would have thought that these people who lived so much on the water could swim; but no, not one of them could swim a stroke, and if they fell into the water they sank like stones. Later on we had a very sad proof of this. It is, however, by no means unusual for seafaring people to be unable to swim; in support of this I need only mention our own fishermen at home.

It had for some time past been decided that our faithful companion, Talurnakto, should accompany us further west, and afterwards possibly go home with us. Talurnakto felt as if he were half a white man, and considered himself quite above his kinsmen; there were few among them with whom he would condescend to converse. But now, as the ice began to get thin and the time of departure approached, Talurnakto seemed to have some misgivings. He had fits of melancholy, and would remain sitting for hours at a time. At first I did not take much notice of it; I thought it was probably a little love-sickness, or something of the sort; but one day, when he was sitting in the cabin, he burst into tears, and, seeing that it must be something more serious, I asked him what it was. He sobbed bitterly, and it was a long time before I could get anything coherent from him. Finally I found out that he did not want to come with us to "Kabluna nuna"—the land of the white men—as they might kill him. I assured him that we were perfect angels, who would do everything that was good for him, but it was of no avail. He would not be convinced, and pointed to some of the pictures of the Boer War. At length I told him that if he did not wish to accompany us we could not and would not force him to do so. At this he was considerably relieved, and his old glad smile again returned. He evidently thought we had him altogether in our power—that we owned him, body and soul; thus the eighth member of the expedition was lost. As, however, I considered it important to have an Eskimo with us for the rest of the voyage, I made it known that Talurnakto's place was vacant. Shortly afterwards the vacant post was applied for by Tonnich, the stout young fellow who was charged with the relief expedition to Lindström. He was evidently very anxious to see more southerly regions. Now, Tonnich was not exactly the man I would have preferred; he was heavy, clumsy, and slow in his movements, but as no others applied and he was always in very good humour, even for an Eskimo, and his conduct had been exemplary, his application was accepted. It was a very solemn moment when Tonnich set foot on board as eighth man. A bucket of water, scrubbing brush, and soap awaited him. To become a

TONNICH.

white man it was absolutely necessary to be thoroughly washed. Wiik supervised the cleaning operation on the deck, so I was sure it was thoroughly done. His hair was clipped and combed and dusted over with insecticide powder, and after that he was clad in normal under-clothing. He now considered that the preparations were completed, and wanted to go at once and submit himself for inspection, but he was given to understand that this was not the usual walking costume, and he was then garbed in a brand-new suit of frieze, made in the prison at home. No one ever felt prouder of their clothes than Tonnich; he laughed with joy, stroked the clothes, and examined himself at all points.

A few days afterwards Talurnakto and two others came with no less than seventy fine salmon, weighing from six to eleven pounds each; in fact, one weighed over seventeen pounds. They were caught at Navyato, and were quite fresh. They were a splendid addition to our stores for the rest of the voyage, the only difficulty being how best to preserve them. It would be too monotonous to be eating salted salmon everyday, so Lund proposed to smoke them. Smoked salmon! the very idea made our mouths water, and Lund's proposal was received with acclamation. He immediately started "The Ogchoktu Smoked Fish Factory"; all he needed was a supply of empty barrels and cases, A number of Eskimo boys were engaged to fetch fuel. A very small kind of heather grew about here in places, and the boys were sent to collect some. They returned with several sacks full, and smoke was soon issuing from the factory chimney. It was not long before the first sample salmon was ready to be served. It surpassed our utmost expectations. The smoked salmon was excellent, and no better could have been obtained from any first-class curing establishment at home.

On July 23rd we met with an accident which might have had very serious results. It was Sunday, and Lindström was about to prepare the midday meal. For this he had to light the "Primus" stove, and was kneeling over the apparatus as usual to get a better purchase on the pumping arrangement, so that he had his face quite close down to the burner; just as he had got it almost in order, it suddenly exploded in his face. He had the presence of mind to seize the apparatus and throw it out. It fell through the open hatchway right down into the hold, but fortunately it was empty. Those in the after-cabin heard a great noise and rushed out; matters looked very bad as the flames burst out of the galley door, but the water hose was lying close by on the deck, and so the fire was put out in a minute. No injury was done to the galley, but poor Lindström got badly scorched. I was sitting in the fore-cabin chatting with Lund and we heard a little noise, but not more than I often heard when the boys were wrestling. Suddenly Lindström rushed in with his face all red and swollen. It flashed on me that there had been a battle for the first time, and that Lindström had had a good thrashing. He was so excited that it was some minutes before he could say "Fire!" Lund and I rushed out, but it was all extinguished. Lindström's face and hands were badly burned. He was rubbed over with egg unguent, but the pain nearly drove him out of his senses. Later, however, in the evening, it eased a little, and the Lieutenant applied bandages. The next day Lindström's good humour had returned, but his appearance was dreadful; he looked like a drunken, dissipated fellow, and was nicknamed "Biffen" (beef); it was almost a fortnight before he looked himself again.

During this time Hansen had to take over the duties of cook, and he performed them very satisfactorily. In baking bread, indeed, he excelled the chief cook. The fact was that whatever Hansen did he did thoroughly, and when Hansen kneaded dough he did it with such energy that you feared both for the dough and the trough, although the result after the baking was finished was the most delicious pastry. Hansen cooked all his dishes to the accompaniment of music; he fried croquettes to the tune of "Vikingebalken" from "Frithjof's Saga," which was, indeed, his only source; cutlets were cooked to the accompaniment of "Isfarten," and blood pudding to "Kong Rings Död," so you could guess what the menu was though you were some distance away on shore.

Lieutenant Hansen and Talurnakto took a trip to Pfeffer River, situated about twenty miles to the west on King William Land, to collect fossils. The channels were now so wide that we could row a boat along them.

On July 28th, for the first time this summer, the harbour was free from ice. Out in the straits we saw that the ice had a bluish tinge, but no cracks were visible. In the previous summer the rivers had been exceptionally full owing to the great quantities of snow.

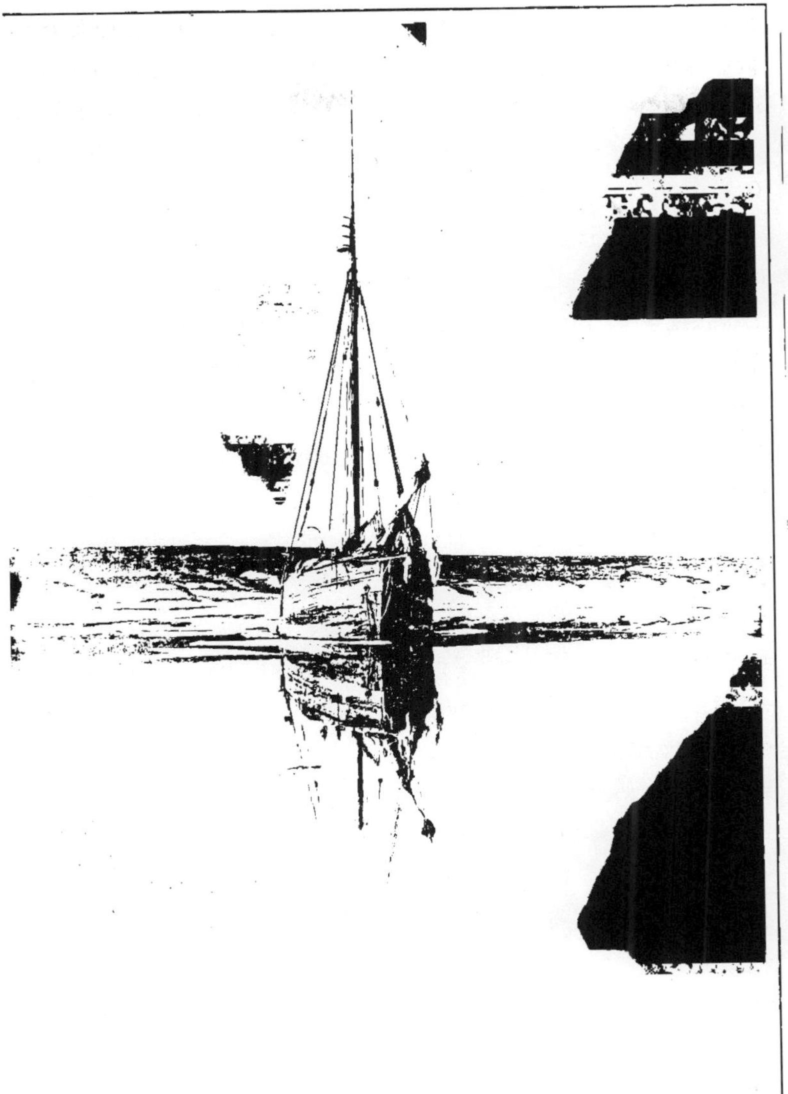

THE "GIÖA" IN SUMMER. GIÖAHAVN, KING WILLIAM LAND

This year they flowed quiet and still, and exerted hardly any influence on the breaking up of the ice, so we had to depend on the sun and wind. There was hardly any current; but for a long time we had a scorching sun every day, and the prospects were rather bright. Towards the end of July the heat ceased definitely, but now the wind came to our assistance. On the night of July 31st a breeze blew up from the north-east with squalls, and sleet fell so heavily that the whole land was white. We had been very anxious about our departure for some time, and were looking out for this north wind with a good deal of excitement. The best of the summer was now over, and the nights, the worst enemy to our further progress, had begun to be noticeable. The ice out in Simpson Strait had up to the present kept exceptionally quiet; no channels had formed, and the ice seemed just as compact and impenetrable as it had been the whole winter. The bluish tinge was, however, a sure sign that it would not need much force to break it up. The only spot of open water was outside Ristvedt River, which was like a bay cut into the ice, and here the gale would get a good hold and begin the breaking up in earnest. And so it happened; in the course of a few days the north-east wind worked wonders. The ice went over to the south, and large channels opened in it in many directions.

We were now free to set sail. With the exception of the meteorological instruments and the dogs, which had to remain on the land till the last moment, everything was on board. The hold was almost completely filled with all our collections. Our most important belongings stood in the main hatchway. First and foremost were the two large re-soldered iron tanks in which were preserved all the observations we had made during these two years. They were so arranged that if they were thrown overboard they would float; both had been marked with the name of the ship. Round these we arranged stores for fourteen days, as well as ammunition and other articles packed in small cases, ready for removal in the event of our being compelled to leave the ship. Here also each of us had his sack of waterproof and other clothing as well as such necessaries as would be required under the circumstances. All our boats and canvas-covered kayaks were perfectly arranged ready to stand a gale. We had made the necessary preparations for sounding our way as we went, till we came out on the other side. We had further prepared three hand-sounding lines, and I arranged a patent pulley on the anchor chocks over which the lead-line ran easily. We had previously distributed the watches as follows: one man at the helm, one in the crow's nest and one at the engine. We deck hands had to arrange matters so that three of us were on deck whilst the fourth took his rest. The engineers took watch in turn and the cook gave us a helping hand whenever he could. We all knew that we were going to have a rough time of it, but the splendid relations which had always existed between us so strongly united us that although we were only seven, we were not easily discouraged.

From "Axel Steen's Hill" we had a splendid view to the west over the Strait, and I went up there two or three times a day during the next fortnight. On August 12th we again got a fresh northerly breeze and realised that if we were to get off, we must take advantage of this. Lieutenant Hansen, Lund, and myself went up to Steen's Hill in the morning. The ice which up to the present had held fast to King William Land along the coast from Booth's Point had now let go its hold, and the channels were open. At 4 o'clock in the evening we were up there again. The ice was still lying around Todd Islands but we thought we could see open water beyond them.

The time had now come and we must make an effort. The departure was fixed for 3 o'clock the next morning; the last preparations were made, the dogs were put on board and after the observations at 9 P.M. the meteorological instruments were also brought on board.

It was with a very peculiar sensation that I went on board for the last time. There was undoubtedly much sorrow mingled with my gladness at leaving. Thanks to my comrades I left Gjöahavn with nothing but happy memories. We had never had a misunderstanding or dispute of any kind. And now as I look back on all that long period I inwardly recall good humour, song, and laughter, and my memories are consequently associated with feelings of gratitude to my comrades for the pleasant days passed at Gjöahavn.

CHAPTER X.

THE NORTH WEST PASSAGE.

OF problems connected with Arctic research, the navigation of the passage to the north of the American Continent has been by far the most interesting to humanity. More lives and treasure have been sacrificed in its solution than in that of almost any other problem. As there is, however, a whole library concerning the "North West Passage," I shall content myself with brief reference to it rather than weary my readers with a historical essay on the subject. I will confine myself to mentioning those voyages and those explorers whose achievements were of the greatest value in the planning and execution of the "Gjöa" Expedition.

John Davis set sail in the year 1585, with the view of discovering the North West Passage. The result was the discovery of the strait between Greenland and Labrador bearing his name. Bylot and Baffin made a fair start in 1616, circumnavigating Baffin's Bay and defining the situation of Lancaster Strait. Dejnev, a Pole, made his way past the north-eastern part of Asia as far back as 1648, and discovered the strait between that continent and America. But his discovery did not become very widely known, and it was a Dane, Vitus Behring, who was the first to make his way through the same strait in 1728, and who had the real credit of discovering Behring Strait. A good start towards the North West Passage was made by these discoveries, but much still remained to be done. In 1778, Captain James Cook penetrated northwards through Behring Strait, and discovered Icy Cape. After this the problem was allowed to rest for a number of years, until attacked again in 1817 by the able captain of an English whaler, William Scoresby, Junior. He was of opinion that the state of the ice had improved sufficiently to warrant fresh attempts. It was thus that John Ross, a Captain in the English Navy, opened the nineteenth century campaign to conquer the North West Passage. In 1743 the English Government had offered a reward of £20,000 for the solution of the problem, and now it renewed its promise. John Ross left in 1818 with the sailing vessels "Isabella" and "Alexander," but fortune did not smile on him. He sailed round Baffin's Bay, passed Smith Sound, and then stood off to the south. At the entrance to Lancaster Sound he suddenly turned homewards. He insisted that the Sound did not exist, and that it was merely a bay. The mountains which he thought he sighted at the inner side of this bay he christened the Croker Mountains. As, however, all his officers refuted his assertions, and maintained that there was a channel, Edward Parry, the capable chief officer of Captain Ross, was sent out in the following year. He not only proved the existence of the Sound, but made his way a long distance westwards, wintering with his two ships, the "Hecla" and "Griper," at Melville Island. This was a giant stride, and the name of Parry must be recorded among the foremost in the history of the North West Passage. John Ross, meanwhile, had not lost heart. In 1829 he again went northwards with the "Victory," a paddle-steamer. This was the first time a steamer was used in the Arctic Ocean. It is, however, needless to say that with large paddle-boxes it was impossible to make much progress in the ice. John Ross passed four winters on the eastern side of Boothia Felix, and was finally compelled to get back in boats, as his vessel was crushed in the ice. Very good results were obtained by this expedition. In later years, his nephew, James Clark Ross, the celebrated Polar explorer, found and determined the position of the Magnetic Pole. Our knowledge of the geography of these regions was also considerably extended, and John Ross regained, in a great measure, his lost reputation. The greater portion of the North American coast was mapped out by means of expeditions in boats, particularly by Franklin in 1819–1822 and 1825–1827. Dease and Simpson continued the work in 1837–1839. The whole of the North American coast was thus, in the main, known, but the North West Passage had not yet been discovered.

Franklin left England in 1845 with the "Erebus" and the "Terror," and favourable results were confidently looked for. Franklin had, during his two previous expeditions, shown such signal capacity that success seemed certain. But, as we know too well, these hopes were not to be realised. Not a single man of the 134 members of the Franklin Expedition ever returned. The uncertainty of Franklin's fate became, during the following years, a burning question to the whole world, and many relief and search expeditions were sent out. Many of these did good work; but the expeditions of Admiral Sir Richard Collinson and Dr. John Rae, especially, were the most important steps towards the final achievement of the navigation of the North West Passage. Admiral Collinson sailed in 1850, on the "Enterprise," into Behring Strait and examined the West Coast of Prince Albert Land and Wollaston Land, where he passed the winter. The following year he proceeded through Dolphin and Union Strait into Coronation Gulf and onwards through Dease Strait, where he was again compelled to winter, in Cambridge Bay, on the south coast of Victoria Land. His soundings and survey of this narrow and foul channel were very helpful to the "Gjöa" Expedition. Sir Richard Collinson appears to me to have been one of the most capable and enterprising sailors the world has ever produced. He guided his great, heavy vessel into waters that hardly afforded sufficient room for the tiny "Gjöa." But, better still, he brought her safely home. His recompense for the heroism shown was, however, but scant. His second in command, Sir Robert M'Clure, who had had to abandon his vessel, the "Investigator," in Mercy Bay, on the north-east coast of Bank's Land, and who was then helped home by others, received all the honour, and one-half of the promised reward went to him and his men as discoverers of the North West Passage. Both of these expeditions were of the greatest importance as a guide to the navigation of the passage. M'Clure had proved that it was impracticable to make the passage by the route he tried. To Collinson belonged the still greater merit of pointing out a really practicable way for vessels—as far as he reached. In other words, M'Clure found a North West Passage which was not navigable; Collinson found one which was practicable, although not suitable for ordinary navigation.

Dr. John Rae was one of the Hudson Bay Company's medical officers. He deserves great credit for his exploration of North Eastern America. His work was of incalculable value to the "Gjöa" Expedition. He discovered Rae Strait, which separates King William Land from the mainland. In all probability the passage through this strait is the only navigable route for the voyage round the north coast of America. This is the only passage which is free from the destructive pack-ice. The distinguished Arctic explorer, Admiral Sir Leopold M'Clintock, pointed out this passage in his report on the "Fox" Expedition in 1857—59, and proved that if the North West Passage were ever to be accomplished, it would be through this channel. I followed the advice of this experienced sailor and had no reason to regret it.

Precisely at 3 A.M. on August 13th, 1905, the windlass played a lively tune on the deck of the "Gjöa." The weather was not of the finest—thick fog and a light contrary breeze. We therefore set the motor going full speed ahead when leaving the harbour. The Eskimo had assembled in the early morning on shore to wish us a last "Manik-tu-mi!" Talurnakto accompanied us out towards Fram Point, and we could hear him calling out his "God-da! God-da!" (good-day) long after he was lost in the fog.

We jumped, so to speak, right into the same doubtful navigation as before, impenetrable fog, no compass, and a very changeable breeze, which was therefore a poor guide. The lead was thrown continually. I put Hansen and Lund on the look-out in the crow's nest, they being the best qualified men for the job, for the cards had to be played judiciously in this game if one wished to come out a winner. The Lieutenant and I myself took the helm in turn, from which point we could better survey the route. Ristvedt and Wiik looked after the engine. The man attending to the soundings had his full share of work; the lead flew up and down so rapidly that it was almost a wonder it did not melt. Ten fathoms and a clay bottom, was the report; then again, eight fathoms, stone; ten fathoms and clay. The bottom along Simpson Strait, off King William Land, was level; sand and clay alternated, and the depth was uniformly about ten fathoms. In this manner we groped our way as far as Booth Point, where we were compelled to stop, as we could not see our way clear to get through the ice, large quantities of which were drifting in an easterly direction. We anchored to leeward of a low rock outside the point, where we were sheltered from the drifting ice. Now and again the fog lifted a bit and we could see Todd Islands ahead of us, surrounded by plenty of ice. To the west of this group of islands we could see open water, and the point was to reach it. At 3 o'clock in the afternoon the fog quite cleared, so that we could gauge our position properly. We were not far from Todd Islands, consisting of three very low islets, large enough, however, to collect a quantity of ice. It did not look very promising from the masthead. There was, it is true, a strip of open water between the bulk of the ice and the most distant island; but it was not reasonable to suppose that this narrow channel extended far inwards, as the ice was drifting at a great speed eastwards, and was probably lying against the western side of the island. However, the best way was to go and see. The weather had meantime become splendid, brilliantly clear, and practically a dead calm. As we advanced and were able to survey the southern point of the island, our anxiety was increased as to whether the ice was lying close up to the western side or not. A channel—so narrow that at a distance it seemed barely to afford room for a rowing boat to pass—was all the open space between the main pack and the island. Then it was a question whether the channel was deep enough. Everything depended upon the configuration of the island. "I think we shall get through," Lund called out from the crow's nest. "I notice stones at the bottom, but we can go close to the shore." This was precisely what we had to do, to squeeze through. Fortunately the west coast of the island was perpendicular, with no shallow bottom near it. But it was only a margin of a few inches, compared with the "Gjöa's" beam, that prevented us from getting stuck. We all heaved a sigh of relief when we had open water ahead of us, to the west.

The Lieutenant and Helmer Hansen, when making their boat trip in 1904, had found two skeletons above ground at Hall Point. These were skeletons of white men; two, no doubt, of Franklin's companions. They buried the remains and built a cairn over their grave. We passed the point just as the sun was setting, and with our colours flying in honour of the dead we went by the grave in solemn silence; the sky and the land then glowing with a soft red, golden light. Our victorious little "Gjöa" was honouring her unfortunate predecessors.

When I came on deck at 2 o'clock next morning we were abreast of Douglas Bay. Tonnich, who knew the locality, gave us the names of the various prominences on the land. He had also noticed the camp where all our Eskimo stayed. It was Kamiglu, a little elevation of about one hundred feet. The tents stood out against the sky, and we could also see the outline of the flagstaff and the little flag. As heavy clouds of fog were now rolling in and becoming very dense over the narrowest part of Simpson Strait, between Eta Island and the land, we made a straight course for Kamiglu. The bottom near the mainland was very uneven, and we therefore cast anchor a good distance from land. As the fog was thickening round our vessel, we started blowing the foghorn at intervals in order to attract the attention of the Eskimo. And soon a kayak shot out of the fog and a hearty "Manik-tu-mi" greeted us. It was Nulieiu, and he was soon followed by others. They were all pleased to see us again, and Lund and I jumped in the dory to accompany them ashore. The fog was no obstacle to the Eskimo. They laughed at us when we asked if they could find their way, and they set off at full speed. Although we had about three-quarters of an hour to row, we went straight to their landing-place. For these people to have been able to row a straight course with such extraordinary precision, without a glimmer of daylight, seems to show that they must be possessed of a sixth sense.

The fog was not so dense inland. Kamiglu is a peninsula with almost perpendicular sides all round, and is only connected with the mainland by a narrow neck between lagoons, both from the east and from the west. Our friends had their habitations in seven tents on the top, in a perfect Arctic paradise. Down below, in the lagoons, they caught all the fish required, and there were great herds of reindeer round the large lakes on the plains. They had killed many and had plenty of meat;

ICHYUACHTORVIK ESKIMO IN HIS KAYAK.

but most of it was kept in depôts out in the open. They were quite willing to fetch it, but as it would take several hours, we abandoned the project and contented ourselves with that obtainable in the camp. We went round bidding good-bye to our old friends; it might be a long while before we met again. At the same time we collected all the meat and dried salmon we could get. Standing

AN ESKIMO FERRY (KAMIGLU, 1905).

outside Umiktuallu's tent was his foster-son, Maniracha, or Manni as we called him for short. He was attired in a pair of sealskin trousers and a greasy old jacket he had got in exchange for something on board. He looked, by the way, as if he had forgotten his morning's ablution; that, however, might happen to anybody. Several months ago Manni asked me, in Ogchoktu, if he might be allowed to accompany me back to the land of the white men. I took his request at that time merely as a sign of his courage, without any serious meaning. However, it seemed to be otherwise. He had just got to know that Tonnich was with us and that he himself was thus shut out; and now he was standing outside the tent crying bitterly. I approached him and asked if he really was so keen upon accompanying us; the tears were simply streaming down the boy's face, and I was really sorry for him when he assured me that this was his only desire in life. Besides, I was ashamed of breaking my promise and felt annoyed; in fact, I would much rather have had Manni along with me than Tonnich. Manni was a quick and smart young fellow and not over seventeen years of age. In this quandary I told him that he might accompany me on board for the present, and I would then see what could be done for him.

At 7 o'clock in the morning the fog began to lift, and I thought it better to get on board and proceed further, as soon as the weather got clear enough. We rowed back to our ship in the dory with thirty-six magnificent joints of reindeer and a large quantity of dried salmon. The "Owl," Manni and two more Eskimo accompanied us, besides four others in their kayaks. Towards 8 o'clock we got on board, the sky being then nearly clear. Only over Eta was it still foggy. I settled with the Eskimo and paid them in ammunition for the meat and salmon. After that I consulted with my shipmates on the subject of Manni. We all agreed upon taking him with us rather than Tonnich, who, on arrival on board, had been dubbed by the appropriate nickname of "Pork-Johnny." Mr. Johnny then came before me, and I tackled him thus: "Well, Johnny, is it true that you absolutely wish to accompany us to the land of the Kablunas?" To this he replied immediately and with surprising frankness: "No, not he; he had no wish to do so!" It was really no easy matter to find out the intentions of these people. Three days ago he would have sacrificed half of his life to come along with us, and now, after living on board like a prince, he had changed his mind. I took this surprise-packet with equanimity, as the Manni question was thus settled. Not entirely, however. His foster-father, Umiktuallu, had still to have his say in the matter —the same pleasant foster-father who had, previously, killed another foster-son by stabbing him. I had first to get his permission to take the boy along with me. And, of course, I did not get his consent. Umiktuallu, who had come on board, wanted payment for the boy. A file and an old knife, however, satisfied him, so he did not value his foster-son at a very exorbitant price after all.

By now the fog had entirely disappeared. We took a last cordial farewell of the "Owl" and our other special friends among the Nechilli tribe and set off. We had before us a most lovely clear summer's day, mild, and perfectly calm. Eta was lying right in the middle of Simpson Strait like a giant who wanted to stay our progress. The two sounds that lay between King William Land to the north and the mainland to the south were not wide. As my readers may remember, Lieutenant and Helmer Hansen had ascertained that the Northern Channel was impassable for the "Gjöa." We centred our hopes, therefore, on the southern one. It was not more than three-quarters of a mile wide, and we knew it was foul. We had long talked about this passage and shuddered at the thought of it. Now we had got to it, and we proceeded with the greatest caution. None of us will easily forget that morning watch. I believe this was our most exciting passage. It was getting shallower and shallower up towards the Sound, but our look-out man reported deeper water beyond the reef we had to pass. The lead was used continually, and the man at the helm had no chance of going to sleep. The helm went from board to board the whole time, just as if we were in thick ice, and, though we managed to get through the Eta Strait, I can vouch for it, as it was my turn at the helm, that the passage did not, by any means, resemble a bee-line. The shallowest water we found was three fathoms.

During the lovely afternoon we had more breathing space again; it had become broiling hot and the sea was perfectly calm. Small lumps of ice were pitching and nodding here and there on the water, with a blue-green reflection in the sun. The lead was still going, but not so feverishly as during the forenoon. The man at the helm was standing dozing and at ease. We could now give some attention to Manni, who hitherto had had to look after himself. I handed him over to Ristvedt, who had the afternoon watch looking after the engine, to make him a Kabluna. Considering the quantity of soap and insect-powder utilised in the process, I was convinced that Manni had had a proper cleansing. We had not the heart to cut his long magnificent hair, but it was well combed, and we noticed no life in it afterwards. His get-up became somewhat picturesque: blue stockinette jacket, sealskin knee-breeches, white stockings and the Lieutenant's old low-cut dress-shoes. His head was covered with a light-blue bathing-cap, which I had at some time or another bought at a watering place. He won everybody's heart from the first start. Manni's laughter banished the most surly airs, and he was undoubtedly pleased with himself, too. He had, it is true, reached the paradise of the Eskimo: the place where you eat as much as you can possibly manage to stow away. By the bye, I was somewhat anxious as to the effect of the change of diet, but Manni did not suffer any ill-effects in this respect. He also enjoyed a smoke of tobacco.

During the evening some ice made its appearance from the south, and presently the whole sea to the south was covered. The edge of the pack extended in a north-westerly direction, and compelled us to follow the same course. We kept alongside it all the way, and sighted a great many small islands in between. It was Queen Maud's Sea which was so full of ice. I had hoped to be able to pass the Nordenskjöld Islands on the southern side, and near land. This, however, was out of the question. The ice was so tightly packed that we had to pass on the outside to make progress. Fortunately, the depth of the water did not hamper us; we could find no bottom with the hand-lead. At daybreak, on August 15th, we had before us the large, newly-discovered group of islands extending as far as our sight could reach, in a north to south direction. The position was clear to us. The ice surrounded the whole group, and we could neither get round to the north nor to the south—we had to go straight through. From the aspect of the islands it was obvious that the waters between them were foul and filled with all sorts of nasty things. We had to get through a small stretch of ice extending all along the eastern side of the islands, and about half a mile outside. We took note of the weakest point, went full speed ahead, all holding on tight. Although the "Gjöa" was small she gave some good thumps, and we got through without much trouble. We now had to follow the open channel southwards to find out whether there was a passage further south between the islands. The lead was again used continuously, and we found the depth of water along the east coast to be uniformly thirteen fathoms. The channel now ceased, and branched off in the shape of a narrow sound between some small rocks. The current had probably formed this channel. The passage was not very inviting, but it was our only one, and forward we must go.

As we turned westward, the soundings became alarming, the figures jumped from seventeen to five fathoms, and *vice versâ*. From an even, sandy bottom we came to a ragged, stony one. We were in the midst of a most disconcerting chaos; sharp stones faced us on every side, low-lying rocks of all shapes, and we bungled through zigzag, as if we were drunk. The lead flew up and down, down and up, and the man at the helm had to pay very close attention and keep his eye on the look-out man who jumped about in the crow's nest like a maniac, throwing his arms about for starboard and port respectively, keeping on the move all the time to watch the track. Now I see a big shallow extending from one islet right over to the other. We must get up to it and see. The anchors were clear to drop, should the water be too shallow, and we proceeded at a very slow rate. I was at the helm and kept shuffling my feet out of sheer nervousness. We barely managed to scrape over. In the afternoon things got worse than ever; there was such a lot of stones that it was just like sailing through an uncleared field. Though chary of doing so, I was now compelled to lower a boat and take soundings ahead of us. This required all hands on deck, and it was anything but pleasant to have to do without the five hours' sleep obtainable under normal conditions. But it could not be helped. We crawled along in this manner, and by 6 P.M. we had reached Victoria Strait, leaving the crowd of islands behind us. The Sound we had passed we christened Palander Strait, after the able commander of the "Vega." The islands south thereof were named Nordenskjöld Islands, after the leader of the "Vega" Expedition. The map of the islands made by Lieutenant Hansen proved most accurate.

Victoria Strait was full of ice-floes, but loose enough to enable us to get through. Outside Lind Island it was thick, but we managed to slip through a narrow channel, getting out on the other side, and reaching open water again. As we were setting sail in the morning our gaff snapped. I then decided to seek refuge in Cambridge Bay, so as to get it repaired. Victoria Land was flat and monotonous. The Dease Strait is deep enough if you keep a couple of miles from the coast. There are shallows off all points and turnings.

We anchored on August 17th, at 5 A.M., on the west side of Cape Colborne, and this was a significant day in the history of our Expedition—for we had now sailed the "Gjöa" *through the hitherto unsolved link in the North West Passage*. We now felt we had got back again to fairly-known waters, so to speak. A sounding was now and then given on the chart, and we felt much more at ease, knowing that we had waters ahead of us which had been ploughed by a large vessel.

Mount Pelly, mentioned by Collinson, is an excellent landmark and easily recognised. Though not more than 800 feet high, it has a gigantic appearance as it rises in its isolation from the plain. I had made up my mind to row ashore and deposit a report, but it was blowing so hard off shore that I had to abandon my project. We repaired the gaff and carried out a few odd jobs. The remainder of the day we had a rest, which we all sorely needed after our toil. It is well, under such circumstances, to have a cook on whom you can fully rely; we could safely leave the care of the vessel to Lindström, who was as good a sailor as he was a cook.

Next morning, at 3 o'clock, saw us on the move again. Collinson's description of the waters was very helpful to us. He had throughout done excellent and reliable work. The sides of the large islands were steep, and the water, therefore, deep and clear. We passed through the Sound between Finlayson Island and the two small islets, situate at a distance of about a mile. We noticed the reef mentioned by Collinson, about two cable-lengths to the south-west of the island. The sea was entirely free from ice.

Our compass, which, after the passage through Eta Strait, had begun to move again, now became quite lively. But we had, of course, to accept its indications with the greatest caution. Next day we passed Richardson Islands. They are high and steep, and quite rich in vegetation. As islets and rocks fill the openings between them, they are not easy to distinguish from one another. In the afternoon we made for Miles Islands, so as to anchor for the night. But the contrary current and wind compelled us to give it up. There are several more islands in this group than are marked on the map. They are all steep towards the east, gradually sloping down towards the west. We lay to during the night of

Chapter X.

August 20th, near Douglas Island, and proceeded westwards as soon as daylight made its first appearance. It turned out a difficult job to find the narrow sound leading out into the Dolphin and Union Straits, between the small islands situated here and Cape Krusenstern. We therefore proceeded northward, right between Douglas Island and Cape Krusenstern, in order to see if we could locate the opening. We, however, got on to a large shoal, which continued until it barred further progress that way. In other words, we had to proceed southwards to get through. To make quite sure, we decided to cast anchor off Douglas Island and do the necessary survey from there. We anchored to the west of the island, half a mile off the shore, in five fathoms of water, with a stony bottom. Douglas Island is quite tiny and flat. There we found a little driftwood, carried out through the Coppermine River. Some old heaps of stone seemed to indicate that the Eskimo frequent the island. Lieutenant Hansen took the necessary sights, and, though we did not see any opening, we now knew where it was.

To some, perhaps, it may occur that we could very well have done this survey under canvas, and that it was unnecessary to stop and retard our voyage on that account. This may be so, but it must not be forgotten that our position was not quite an ordinary one. Bearing in mind our running aground at Matty Island, we had quite decided not to risk a recurrence of that experience if we could possibly avoid it. We would rather sacrifice a few hours than jeopardise our vessel in these very hazardous waters, with a ragged stone bottom and shallow water under her keel, an unsafe compass, and a small crew. We were, so to speak, standing on the threshold of our goal, attempted unsuccessfully by so many before us, and, taking this into consideration, it was an easy task to restrain our impatience to get along as speedily as possible and out of our difficulties.

At the first sign of daybreak we were at it again. We were compelled to keep southwards to avoid the shoals between the mainland and Douglas Island. The water was now getting deeper. Finding eventually that we had got far enough to the south, we turned off to the west, shaping our course towards the point where we expected to find an opening. It was an exciting time. Fortunately the deep water continued—we found nowhere less than seven fathoms—we neared the mainland without trouble, and found the passage all right. At 3 P.M. we passed Liston and Sutton Islands, and stood off into Dolphin and Union Strait. My relief at having thus got clear of the last difficult hole in the North West Passage was indescribable. I cannot deny that I had felt very nervous during the last few days. The thought that here in these troublesome waters we were running the risk of spoiling the whole of our so far successful enterprise was anything but pleasant, but it was always present to my mind. The whole responsibility for crew and the vessel rested on me, and I could not get rid of the possibility of returning home with the task unperformed. The thought was anything but cheering. My hours of rest and sleep were principally spent, during this time, in brooding over such thoughts, and they were not very conducive to sleep. All our precautions and everybody's careful attention notwitstanding, any moment might have some surprise in store for us. I could not eat. At every meal-time I felt a devouring hunger, but I was unable to swallow my food. When finally we got out of our scrapes and I regained my usual calm, I had a most rapacious hunger to satisfy, and I would rather not mention what I managed to dispose of.

We could now discontinue the laborious watches of eighteen hours a-day and revert to the normal arrangement of six-hour watches. Barring a few small interruptions in the shape of fog and contrary wind we made fair progress westwards. We did not sight Clerk Island at all, although the weather was clear, and it should have been well within range of vision. Its existence would, therefore, seem somewhat doubtful. We encountered small lots of ice now and then which reminded us that we were in the Arctic regions and must be prepared for eventualities.

On August 26th, at 4 P.M., we sighted a high land to windward. The air was very misty, and as, according to our reckoning, we should be abreast of Cape Parry, I thought this was that we saw. During the early morning the air became clearer, and I knew then that this land was not Cape Parry on the mainland of America, but Nelson Head on Baring Land. The error was not quite insignificant, to be sure. But my misgivings on this head were appeased when told later by American whalers of the ludicrous mistakes they often made in these waters. There is probably a lot of iron in the mountains here, and the compass therefore becomes utterly distracted. Then there are strong currents, and the united influence of these factors may confuse the most conscientious navigator even more than it did when we mistook Nelson Head for Cape Parry. We were, of course, wholly unacquainted with the condition of things. When we had found our bearings, we continued our voyage at full speed, having a fair wind as well as the current right behind us.

At 8 A.M. my watch was finished and I turned in. When I had been asleep some time, I became conscious of a rushing to and fro on deck. Clearly there was something the matter, and I felt a bit annoyed that they should go on like that for the matter of a bear or a seal. It must be something of that kind, surely. But then Lieutenant Hansen came rushing down into the cabin and called out the ever memorable words: "Vessel in sight, sir!" He bolted again immediately, and I was alone.

The North West Passage had been accomplished—my dream from childhood. This very moment it was fulfilled. I had a peculiar sensation in my throat; I was somewhat overworked and tired, and I suppose it was weakness on my part, but I could feel tears coming to my eyes. "Vessel in sight!" The words were magical.

Chapter X.

My home and those dear to me there at once appeared to me as if stretching out their hands—"Vessel in sight!"

I dressed myself in no time. When ready, I stopped a moment before Nansen's portrait on the wall. It seemed as if the picture had come to life, as if he winked at me, nodding, "Just what I thought, my boy!" I nodded back, smiling and happy, and went on deck.

It was a wonderfully fine day. The breeze had veered round somewhat to the east, and with the wind abaft, and all sails set, we made excellent headway. It seemed as if the "Gjöa" understood that the hardest part of the struggle was over, she seemed so wonderfully light in her movements. Nelson Head was a long way off to the north. The flat-topped promontory looked grand in the morning sunshine, melting in the white snow, and throwing dark-blue shadows into the parallel fissures of the mountain side. A heavy, bright swell rocked the vessel pleasantly, and the air was mild and soft. All this was observed in a moment. But it did not arrest our attention for long. The only objects between sky and sea that possessed any interest for us then were the two mastheads on the horizon. All hands had come on deck, and all glasses were levelled at the approaching vessel. All faces were wreathed in smiles. Not much was said; one of the telescopes was lowered——"I wonder——!" And it was raised again. Another one lowered the telescope, and also remarked: "I wonder!" On the appearance of the unknown vessel we hoisted our Norwegian flag. It glided slowly up under the gaff, every eye watching it. Many pleasant words were whispered to the flag, it seemed as if everybody wanted to caress it. It had become a bit worn and ragged, but it bore its wounds with honour.

"I wonder what he'll think when he sees it?"

"He'll think it is a venerable old flag."

"Perhaps he's an American."

"I shouldn't be surprised if he were an Englishman."

"Yes, he will see by the flag what we are!"

"Oh, yes—he will see we are boys from good old Norway!"

The vessels were approaching each other very rapidly.

"There! up goes the American flag," sang out the watchman. He had the long telescope which had been placed on deck. This proved to be correct, and we could now all see the Stars and Stripes under the vessel's gaff. They had seen and recognised our flag by now, that was certain. Dense steam was issuing from the vessel's side; evidently they had a motor, the same as we had, and were advancing rapidly.

It was time now to tidy ourselves a little in preparation for the first meeting. Four of us were to go on board the ship, the other three had to remain on the "Gjöa" and look after our vessel. Our best clothes were hurriedly got out. We dressed ourselves according to our individual taste. Some preferred Eskimo costumes, and others our Norwegian russet. One found that sealskin boots looked best for the occasion, others preferred ordinary sea-boots. We also cleared up on deck as well as we could. The American could certainly scan our deck in every detail, from his crow's nest, through his telescope, and we wanted to make as decent an impression as possible. We were now so near each other that the whole ship was visible from our deck. It was a small, two-masted schooner, painted black; she had a powerful motor, and the foam at her bows was spurting high. She also carried sail. We got the boats clear, hove to, and lowered the dory, the most seaworthy of them. It was certainly not much to look at, and the commander had no easy stern-sheets, with a flag, to sit on. But the boat was in the style of the vessel to which it belonged, and we were not on a pleasure trip. The American had stopped his engine, and was waiting for us. With two men at the oars we were soon alongside of him. A line was thrown down to us; I caught it, and was again linked with civilisation. It did not, however, make its appearance in any great glory.

The "Charles Hanson," of San Francisco, did not seem to be rigged out in a very luxurious manner. A ladder, by-the-bye, was superfluous, as the ship was deep in the water. We took hold of the chain-wales and crawled on board. Our first impression was most peculiar. Every available space on deck was occupied to such an extent that it was nearly impossible to get along. Eskimo women in red dresses, and negroes in the most variegated costumes were mingling together, just as in a land of fable.

An elderly man with a white beard advanced towards me on the quarter-deck. He was newly shaven, and nicely dressed, evidently the master of the ship. "Are you Captain Amundsen?" was his first remark. I was quite surprised to hear that we were known so far away and answered in the affirmative, owning that I was the man. "Is this the first vessel you have met?" the old man asked. And when I admitted it was so, his countenance brightened up. We shook hands long and heartily. "I am exceedingly pleased to be the first one to welcome you on getting through the North West Passage." We were then most courteously invited down below to his cabin. There was not much room, though slightly more than on board our own vessel, the "Gjöa."

Captain James McKenna, the master of the "Charles Hanson," was a man of medium height, corpulent and between fifty and sixty years of age. That he was an old Arctic trader was evident from his looks. The deep wrinkles and copper-coloured face told plainly of cold and murky weather. His personality was jovial and agreeable. He asked if we wanted anything, in which case he was ready to help us to the best of his ability. The only thing we missed so far was news from home. But unfortunately he had none. That is to say he had some old newspapers, but . . . . Old! Yes to you! To us they are certainly absolutely fresh! He brought out a bundle, and by a wonderful coincidence my eye first alighted upon a head-line which made me stare. "War between Norway and Sweden." I swallowed the article in hot haste, but it only gave a moderate amount of information. Captain McKenna had left home long ago and could give no more particulars. We sought further information all over the ship, but no one knew any more about it. This uncertainty was more unsettling than our previous total ignorance, but it could not be helped; we had to put aside our anxiety and wait.

After a very good dinner Lieutenant Hansen and I began culling as much information as possible regarding the navigation ahead of us. McKenna was the Senior of the American whalers and knew the North American Coast better than anyone else. What we prized particularly was the set of American charts for the continuation of our voyage. They were of a more recent date than ours and contained many new items. With marginal notes and indications of courses by the old, experienced captain they were a real treasure to us. They were somewhat worn and tattered, and we, therefore, packed them up most carefully. Then about the condition of the ice. Did he think we could continue in a westerly direction without hindrance? He told us that when inward bound he had been hampered by ice near Herschel Island, but that at the present late period of the season we were hardly likely to meet any obstacles of consequence. We would in any case reach Herschel Island quite easily. He was certain of this, and as he was himself going to winter on that island it might happen that we would meet again. Before going into winter quarters he intended making a trip as far as

Banks' Land to look for whales; so far he had been unlucky and got none. His motor was very powerful, and he would probably catch us up on his return voyage to Herschel Island. In addition he gave us every possible information about the waters ahead of us. It was pleasant to hear that the bottom along the whole coast westwards was even, so that we could navigate safely by the lead. We had not been spoilt by safe navigation, so we looked upon the remainder of our voyage as a mere pleasure trip.

The breeze kept up well, and as I considered I could not afford to lose more of it, we said good-bye to our amiable host after a visit of two hours' duration. When leaving he made us a present of a bag of potatoes and another of onions. As it was a long time since we tasted such luxuries, we gratefully accepted the gifts.

We were awaited on board with eager expectation. For the present we agreed to look with great distrust on the reported war between the two united kingdoms. The potatoes and onions became the centre of joy, most of us being fond of these vegetables. We then dipped our flag, set all sail and continued our voyage. McKenna proceeded eastwards to try his luck.

The next afternoon we passed Franklin Bay. "The smoking rocks," mentioned in several previous reports by travellers, were still active, thick smoke issuing therefrom. The Bay was unfortunately full of ice, so that we were unable to land and inspect the phenomenon more closely. On the outskirts of this ice a bear was standing, evidently watching us with interest. A fit of blood-thirstiness naturally seized our hunters, and two of them fetched out their rifles and took to a boat. When Master Bruin smelt a rat he plunged into the sea and commenced swimming away. He was soon caught up, however, and shot. He was a comparatively small fellow, but the fur was exceptionally beautiful. The great Nimrod himself got the skin, and we gave the flesh to the dogs, who revelled in it. The flesh of the bear is not bad, but we had still plenty of reindeer-meat, so bear did not tempt us. We then went on again, but the hunters soon discovered more game. Two bears were lying on an ice-floe, apparently asleep, as they showed no sign of life. To the undoubted disappointment of our sportsmen I let them alone, however. The wind was fair and we must utilise it. There was plenty of ice, but still so loose that we could force our way through. However, we now had six hours' darkness during the night, and while this lasted we could not possibly keep going with the present state of the ice. As we made fast to the ice, in the evening, it was calm. When daylight appeared, the ice was lying close up to Cape Bathurst, without a fathom of open water. The ice was loose from north-east and also north to east. I thought it better to wait and bide our time. If the breeze from north-west eased off during the day I had good hopes that the ice would drift from the land and make way for us. Fog also made its appearance with the north-west wind, and presently we could not see a yard ahead of us. Later in the day the wind eased off as expected. The fog lifted, and by 5 P.M. we had it very fine and clear. Soon after, the ice began to move, but not in our favour, unfortunately. A great pack was gliding down upon us from the east and threatened to imprison us completely. The ice was still quite loose to the north, but it could not be long before we should feel the pressure of it as it advanced. In order not to be shut in I decided to get steam up, proceed to the north-west, and try if we could not go round the ice that way and get under land again further west. We just managed to squeeze out before the two packs of ice collided, but it had now become so dark that we could discern nothing. We were compelled, therefore, to stop and use our old tactics of lying-to.

Next morning we found that while lying-to we had got a long way westward. But the loss was not very important. We had had the opportunity of ascertaining that we could not get along that way. We would have to try to get back to our old position, where any change must first be felt. The south-easterly breeze became fresher, and we had plenty of hard work beating our way back. But, when we reached our previous position, it turned out, sure enough, that the ice had dispersed, and we were thus able to proceed towards Cape Bathurst. There we found a channel which was not very broad, but being able to get close in shore we managed to get through. The coast here is a bold one.

At 5 A.M. we passed two American whalers, who had their boats out looking for whales. We did not want to disturb them in their work, and, besides, we had no particular interest in talking to them. We therefore passed them by. It was my intention to go as far as Bailey Island, fill our water tanks at our ease, and pump petroleum from the fore tanks into the engine tanks. As, however, the wind was fresh from the south-east, and we made good headway, I thought it just as well to proceed. When passing Cape Bathurst we perceived a large number of Eskimo on shore. They waved and made signs to us, and even hoisted a flag. A large wooden house was to us a greeting from civilisation.

At 4 P.M. on August 30th we passed Bailey Island, with a strong, fair wind, without sighting any ice. McKenna advised us to keep near land all the time, but, as things were shaping now, it was too tempting a prospect to proceed right across to Herschel Island, and thus shorten the voyage. This we accordingly did.

Off Cape Bathurst we encountered the muddy, brown water which the Mackenzie River throws out. It is of no use to watch from the crow's nest, here, for shallows through these muddy waters, and the lead was our only guide. The bottom was even, however. The wind increased, and caused the sea to become a bit rough; and it was no ordinary sea, with blue billows and white foam on the surface; these billows were brown and the foam yellow, full of sand and gravel. The chart indicated a shallow of three and a half fathoms in this neighbourhood, and we did not care to come into contact with it in this kind of weather. The position became a good deal worse, owing to numerous large masses of scattered drift ice, which it was difficult to steer clear of in the dark. All went well, however, and at daybreak next morning everything was all right. The breeze lasted all day. The weather, however, was dull, and we had to admit that McKenna was right, as we suddenly ran up against the big pack, rendering further progress impossible. We kept a southerly course along the edge of the ice. The depth was about eighteen fathoms, but decreased further south. When, at 8 P.M., we had reached eight fathoms, and it became pitch dark, we lay to till daybreak. During the night we found five fathoms of water. At 4 A.M. we proceeded westwards at a slow speed, using the lead all the time. The fog was impenetrable, but we had noticed the night before that we had a clear course to the west. However, there were a few islands here, which made us cautious. But, as we still kept at five fathoms after proceeding slowly for half-an-hour, I concluded that we were far enough away from land, and set all sail and put the motor at full speed ahead. We were then getting along splendidly. At 5 o'clock the fog lifted a bit, and we sighted an island about a mile to the south. This must be Hooper Island. At the same time we saw two barques, which were waiting their chance. Soon afterwards it became very dull again, with snow from the east. The "Gjöa" made good speed, the foam spurting off her. We used our fog-horn, hooting and tooting in honour of the vessels coming along. We passed several points covered with ice, which compelled us to keep a southerly course towards the Mackenzie River.

At 11 A.M. it cleared again, and we then sighted two barques a long way behind us. They were making the same course as we and soon overtook us. They turned out to be the "Alexander" and the "Bowhead," of San Francisco, commanded by Captain Tilton and Captain Cook. The "Bowhead" was an old acquaintance. She had been bought from Norway, being previously used for many years in the Arctic trade, and known as the "Haardraade." Both vessels hailed us and proffered every assistance. We did not, however, need any, so we thanked them for their kind offers and let them pass. They informed us that they were now leaving the ice, homeward bound. But they were to call at Herschel Island first, and we mutually expressed our hopes that we might meet again there in a couple of days' time at the most. Our hopes were not, however, fulfilled quite so soon as we then thought they would be.

During the night before September 2nd we again had a most unpleasant time of it, with four fathoms of water, plenty of ice, and pitch dark. We lay to as well as we could, and in the morning we worked ourselves forward to reach the open channel. This was not very wide, but there was, fortunately, plenty of water close to the shore. By 2 P.M. the wind increased with squalls, and we were going at a great pace. Lund was in the crow's nest and I was at the helm. We carried full sail; in the smooth sea we were not hampered by big waves, and we had never made brisker headway with the "Gjöa." However, one had to be quick with the helm when surrounded by ice, and when we got into the channel, in the course of two hours and a-half, I could feel the effect of my turn at the helm. My hands were full of blisters, and my clothes were dripping wet from perspiration. The lead was going the whole of the time. Close to the shore we found two fathoms of water. As the channel was too narrow to beat to the windward, all we could do was to make fast to the solid ice. We put out two ice grapnels. The land near which we were moored was Cape Sabine, about sixty feet high, steep towards the sea and flat on the top. We left the cook in possession and went ashore. The beach was quite covered with driftwood; enormous quantities were piled up, and trunks fifty feet in length were by no means rare. All this wood seemed most glorious to us. Our gaff had been broken and repaired so many times that it was of no further use, and here we had plenty of material for gaffs. We separated along the shore and each of us looked about for the best pieces. There was enough to choose from. Even here on the hill, facing north, where no sunshine penetrates, except for a short time at night, we found some beautiful flowers. We collected some large bunches of forget-me-nots and other varieties for decorations on board. Then we walked up to the top of the hill to have a look further inland. Long, billowy fields, covered with high grass, extended so far as we could see, and in the valleys there were bushes exceeding the height of a man. To us this looked like a paradise. A few ducks were lying off the shore, but, to the disappointment of our hunters, we did not discover any other animal life. We proceeded a little further inland and hit upon a few discarded Eskimo huts. A couple of old sledges and bows and a rusty barrel of a muzzle-loader were articles that might have been left here by either white men or by Eskimo.

Next morning the condition of the ice had not materially improved. But the wind had abated sufficiently to allow us to proceed, by the aid of our motor, through the narrow channel. We fetched on board the pole intended for a new gaff, let go, and departed. About an hour later our look-out reported from the crow's nest that a boat was approaching us from land. At first we thought it was an Eskimo boat, but soon discovered it was manned by two white men and one Eskimo. We took them on board, and, curiously enough, the first of the men addressed us in Norwegian. He was a Norwegian, named Christian Sten, who had been second mate on board the schooner "Bonanza," of San Francisco. The schooner left home simultaneously with us, and, like ourselves, had passed the winter in these regions. The vessel had, however, been damaged by ice and by stranding, and a few days ago they were compelled to run her ashore at King Point to save her from sinking. Mr. Sten was now staying ashore, with one of the vessel's harpooners and some Eskimo, to keep watch on the ship's provisions and other equipment. Captain Mogg,

"Bonanza.' "Gjöa." Our Residence.

THE WRECKED WHALER "BONANZA" AT KING POINT.

the commander, had gone with the remainder of the crew to Herschel Island, in boats, to find some means of getting southwards to San Francisco on board another ship. We could now see the wreck under the bold cape before us. Mr. Sten told us the ice was lying close to King Point, and that for the present we could get no farther. He did not doubt, however, but that the ice would slacken. He had seen it break up as late as October 9th.

We arrived at noon, and found the state of the ice as described by Sten. We approached a large sheet of solid ice lying outside the wreck, and made fast to it. Little did we dream then that King Point was to be our residence for the next ten months. We rowed ashore to have a look at the "Bonanza" and at Sten's little colony. Captain Tilton, of the "Alexander," was the oldest commander of the company to which the "Bonanza" also belonged, and, when sailing past, two days previously, he had given instructions to Sten to assist us with anything we might need. We were well provided, but, as such a friendly offer was made us, we profited by it to obtain a few things we could do with. We changed some canned provisions, as we wanted to try the American varieties, and Sten, on the other hand, wished to try the Norwegian. We obtained various other little articles, and I wish here to express my sincere gratitude for all the help afforded us by Sten and the "Bonanza." Sten had spent many winters on the North-American coast, and was in a position to give us much useful information about the country and the fairway. He also knew the Eskimo living here, which was of great importance.

Manni had by this time accustomed himself to the life on board; he was dressed entirely as a Kabluna, and, as he was an exceptionally clever huntsman, I had made him a present of a carbine and a shot gun, of which he was very proud; he looked after them most carefully. I asked him if he would now like to leave us and go ashore, but he positively declined. I took him with me to visit the Eskimo staying with Sten, and, as it turned out, they easily understood each other. A word now and then might differ, but, taken altogether, their language was identical. These Eskimo, one man and three women, hailed from the Kotzebue Sound locality, near Behring Strait; they had come there with the whalers. They called themselves Nunatarmiun Eskimo. The inhabitants on this coast called themselves Kagmallik Eskimo; but civilisation had had its corrupting influence on them, so that, instead of several hundred families, their number was reduced to a handful. The Kagmallik Eskimo were taller and of a finer build than the Nunatarmiun.

Sten was busy building a house for himself on a small spot on the slope close to the provision and other stores that had been landed. We also paid a visit on board the "Bonanza." She had capsized near the land. The foremast had been cut away, but the mainmast was still standing. A hawser was stretched from this to the same ice where we had run one ashore. The hold was full of water, and a quantity of casks and barrels were floating about. Much material had been cut out of the vessel. With permission, we also took what we required. especially cordage, blocks, lanterns, &c. We also accepted, with pleasure, a small stove. Should we have to pass another winter here, which now appeared likely, it would come in very handy. Certainly, we had plenty of material to build ovens with, and a blacksmith capable of doing it. However, it was better to get one ready made and save the labour. It suited our palates also to get a change of diet. The canned American fruit, especially, was a great success on board the "Gjöa." We, also, were useful to Sten. He had a lot of work to do and needed assistance. He might, of course, get it later on from the Eskimo, but it was preferable to get it over before snow commenced to fall.

We were not the only ones waiting for a change in the condition of the ice. A large number of Eskimo, who had left Herschel Island in boats for the Mackenzie River, were held up by the ice about four miles west of us. From the top of King Point we could see the rigging of a schooner, in the direction of Key Point, fifteen miles to the west. This vessel belonged to the Eskimo. They had bought it in exchange for furs and used it as a whaler. They had now run her ashore. However, they got her off before the ice closed up, and succeeded in reaching Herschel Island. The Eskimo living here are capable seamen, whalers, etc., and the

Americans do not therefore bring a large crew, as they can find plenty of hands on the spot who are both more capable and manageable than their own.

Ristvedt and Manni were out hunting and brought home a great many ptarmigan. The Lieutenant and I set nets and procured many a meal of fresh fish: Lund was

ESKIMO AT KING POINT.

working like a nigger to finish the new gaff, ready for our departure. Wiik and Hansen, at my request, volunteered to help Sten in building the house. The point was to get the roof on before the advent of the snow. Their help was an acquisition, and they also appreciated the change of duties, and more particularly diet. The days passed without any noticeable change in the ice. We had to accustom ourselves to the probability of remaining here for the winter. Our main concern was whether our anchorage was a safe one. The bay outside was very shallow and full of ice firmly planted on the bottom, so there was little prospect of being much squeezed. Sten had also told us that three whalers had wintered here without discovering any movement of the ice. The channel was still open to the east, so that we could get out to Shingle Point, fifteen miles farther east, where there was said to be a small harbour. But this was rather uncertain, and as we had company and help on the spot, we decided to remain where we were.

New ice, several inches thick, was now forming every night, and our fate was soon sealed for another winter. On Saturday, September 9th, we were able to walk on the ice, and we must therefore regard this as the opening chapter of our third winter.

CHAPTER XI.

THE THIRD WINTER.

ON the very day the ice was strong enough to bear, we received our first visitor. It was Mr. Fraser, a missionary, coming from Herschel Island and bound to Fort McPherson, the Hudson Bay Company's most northern station on the Mackenzie River. As conductor he brought with him an Eskimo, named Roksi. They had had to stop on account of the ice, and were now living in a tent on the shore, about four miles west of us. He reported that five vessels had been shut in by the ice in the harbour on Herschel Island. There were six other vessels to the east of us, exact positions uncertain. Thus no less than twelve vessels were here in the ice, and only three of them were prepared for wintering. This did not sound very promising. Roksi was a Kagmallik, and, as his father had been a chief, he considered himself a most important person. His kinsmen, however, were by no means overawed by his pretended noble birth, and only laughed at him. These Eskimo had the nasty custom of punching holes in their under lips, at the corners of the mouth, and of inserting therein a pair of large bone buttons by way of ornament:

The more civilised among them had, however, taken off the buttons. The holes would then contract and form nasty scars.

On Monday, September 11th, we started building ourselves a house. This winter we were going to have two, both constructed of driftwood. One was for us to live in, the other to serve as an observatory for the

CHR. STEN. ROKSI.

magnetic variation instruments. Our residence was to contain two rooms, a bedroom for four men and a combined kitchen and dining-room. All of us preferred living ashore. To avoid the humidity on board, I found it answered best to remove the entire cooking department ashore. Besides, our cook and Sten had become intimate friends, and we wished to profit by this as much as possible. Sten was, I may say, a splendid cook. He had an excellent large oven in his house, which was now completed, and he could prepare some wonderful dishes. The Lieutenant and I were to remain on board together with Manni, to look after the ship. The architect and the smith undertook the erection of our residence. They decided to build it on the model of a Lapp turf-hut, this being the most practical form. As assistants they had Hansen and Wiik. The most level spot on the slope was chosen for the site. The whole morning was spent in collecting materials. By the afternoon the two long walls had been completed. Meanwhile the Lieutenant and I made ourselves as comfortable as possible on board. The oven, which we had annexed from the "Bonanza," was erected, and this supplied us with all the warmth we had so missed during the past winters. We had, however, to saw and chop wood for the oven, but it did not take long to teach Manni this work. It was as well he should have something to attend to. His duties were furthermore made to include keeping the cabin clean, *i.e.*, sweeping away most of the dust, etc., every morning. Formerly we had had no time to do this more than once a week, and then it did not look very nice. The Lieutenant and I looked after the fishing. Every morning we made an opening in the ice and pulled up the fish we required. We generally secured between twenty and thirty "whitefish," a species found in abundance along the North American coast. It much resembles a large herring. The Lieutenant had rigged up one of our sailcloth boats on a lake and was hunting ducks successfully. Manni also hunted on shore and brought home geese, ducks, and grouse, so we had plenty of fresh food all the time.

By September 15th our house was roofed in. The two rooms were of about equal size. Bunks for four men had been put up in the inner one. In front of each bunk there was a form, and a small table occupied the middle of the floor. Wiik had a folding table to himself, on which he could evolve magnetic curves. One of our petroleum tanks was used for an oven, prepared by Ristvedt for the purpose. A funnel was made of iron plate obtained from the "Bonanza." The outside room was dining room and kitchen combined. The kitchener was made out of a petroleum tank. Sten supplied us with a plate with six holes in it, as he had a double supply. Out of these articles our clever blacksmith made the most wonderful caboose "that ever was." If the Lieutenant wanted to get something extra special or wonderful made he would go up and see Sten in the kitchen. He had no baking oven; he continued to bake bread on the "Primus" as before, and, as it may interest housewives, I may mention that our brave Lindström made the finest and lightest wheaten loaves imaginable for three successive years in a little oven on the top of the "Primus." To bake eight large loaves required about half a pint of petroleum. At the side of the "caboose" there was a long table where all eight of us could sit down to our meals. A small dresser and a box, to keep sundries in, completed the furniture of the dining saloon. Outside the dining room there was a small entrance hall where one could brush off the snow before walking in. The house was built facing north and south. The floor was made of a few boards we had brought along with us for this purpose. But although the ground was the most level one on the bank, the floor had a slight slant which Lindström especially

OUR RESIDENCE AT KING POINT

complained of. It is true he had to be about there all day. We covered the house inside with sail cloth. The outside was covered with moss. When the snow fell and completely covered us in we should have a fine house. Light was obtained through skylights facing east. A shed made of sail-cloth along the eastern wall served for the storage of wood.

When our house was roofed in, Wiik and Ristvedt

started on the Variation House. The ground for this was selected about two hundred yards from the other house on a small open point terminating abruptly towards the sea, to the north. A tent was at first erected here in which Wiik, through various observations, determined north and south, the direction of the magnetic meridian

THE REGISTER HOUSE AT KING POINT. THE STAKES MARK THE WAY IN A SNOWSTORM.

whereon the house was to be built. There was a long acclivity to the south, terminating in extensive plains. We marked the road so far, by means of poles, in case of possible snowstorms.

During this time I had frequent visits, especially from Eskimo who, like ourselves, were shut in by the ice.

Chapter XI.

The missionary also came now and again. When he decided to abandon his trip to Fort McPherson he came one day, accompanied by Roksi, to say good-bye. I invited them to dinner, and at table, Roksi, who spoke English pretty well, told us that the Eskimo in this neighbourhood had a word for "thanks," viz., "koyenna." The missionary would insist that this word was introduced by the Christian Mission, which Roksi, however, denied. The missionary became somewhat engrossed in the argument, and mentioned that at any rate such words as "Amen" and "Hallelujah" were introduced by the Mission. "Not at all," said Roksi, "we said Amen and Hallelujah long before the Mission came here!" It was said with the greatest assurance, and we all roared with laughter.

After finishing our building operations we covered the vessel with sails and made our final preparations for the winter. This time we made the entrance aft, on the starboard side. A small cabin door from the "Bonanza" was inserted in the sail-cloth, and a big, wide staircase was fitted up in front of it, from the ice, so that we were now fixed up in first-class style.

One day the first large caravan of Eskimo passed us. They were part of those that had got stuck with their boats, and they were now proceeding towards Fort McPherson on sledges. They formed a motley, even festive-looking procession as they came driving along, between the "Gjöa" and the "Bonanza," in their dog-sledges with their merry harness-bells; it reminded us

(*Photograph taken by Moonlight.*)

IN WINTER QUARTERS AT KING POINT.

somewhat of our Christmas sledge-parties in Norway. These Eskimo drove in a different manner to our friends the Nechilli. The dogs were generally harnessed in single file, but sometimes two abreast, and so that they could not get out of their allotted position. This, of course, had its advantages as well as disadvantages.

Kunak's House.
Sten's House.
THE COLONY AT KING POINT.

Sten had covered his house with turf, and Kunak the Eskimo had completed his house, alongside Sten's. When the winter started with snowstorms in earnest, it found the colony on King Point quite prepared. We were twenty souls, all told, encamped together here for ten months. Lieutenant Hansen, Manni and I stayed on board the "Gjöa." The other five "Gjöa" men lived in our house. Sten's house was situate fifty yards further west. It was built of planks and boards from the "Bonanza" and looked for all the world like a villa residence. It consisted of two rooms. Sten lived in the inner one with his wife, Kataksina, and their little daughter Annie. It was both spacious and comfortable. The outer room was used as kitchen and living room by both families. The harpooner, Jimmy, and his wife, had a large comfortable bed close to the kitchener. It may have been a little warm at times; I refer to the days when the bread was being baked, with a temperature of 112°. But the Eskimo Nein and his wife, who were berthed on a wooden trap door right above the kitchener, were worse off still; it was a wonder to me that they could stand it. However, they were soon going out hunting and would be spending most of the winter in the fields. These two rooms were lighted by a huge skylight in the roof. Outside the kitchen there was a large hall, divided in two, one part serving as workshop and one for storage of provisions. The workshop led into a large, roomy shed built of planks and old sails. A door from this led out into the open. Everything was practically and conveniently arranged.

Kunak had constructed his little cottage so that it joined Sten's. There was only one room, and the furniture consisted of two beds, a table, and an oven. This cannot be considered excessive accommodation, seeing that Kunak had to house his old mother, his wife, and two children. He frequently had visitors,

Kunak.

KUNAK AND HIS FAMILY. SUMMER AT KING POINT.

and then there might be as many as ten persons in his room.

There were about as many dogs as people in our colony.

We collected wood for the winter most assiduously. There was plenty about, and to prevent it being buried

KING POINT. SHORE STREWN WITH DRIFTWOOD.

in snow we piled it up on end in big heaps, and had it carried home as required. An important discovery was made, which was especially welcome to our cook ; the sea-water was quite fresh, and it was excellent in every respect, both for drinking and cooking purposes. This seems strange, certainly, but it is owing to the great Mackenzie River, which is not far off.

Chapter XI.

We were somewhat plagued by colds. Manni, especially, suffered so much from them that we had to keep him in bed for several days at a time, notwithstanding his protests. He also suffered from hæmorrhage from the nose, hardly a day passing without it.

I had long been watching the ice, to see if its condition would soon permit of our reaching Herschel Island, and

THE LAND BETWEEN KING POINT AND KEY POINT.

so making inquiries about the post, which was to leave there in the near future. All of us were, of course, most anxious to get news from home. I had arranged with Sten to accompany me; he wished to go to Herschel Island and have a talk with the whalers. The Eskimo west of us had promised to let us know when the ice would bear. The portion near Key Point did not acquire

THE AMERICAN WHALING FLEET AT HERSCHEL ISLAND, 1905, 1906.

the necessary stability very easily, as a rather large river emptied itself into the sea there.

On Sunday, September 24th, an Eskimo passed on his way to Herschel. If he could get along, we ought to manage it. On the following Tuesday, therefore, we made a start, the first thing in the morning, with a sleigh and a good team of dogs. The road was not very good, as the new snow had not properly set. Every now and then we had to plod through loose drifts. Four miles to the west we came across the first camp of Eskimo, comprising several tents. All their boats had been dragged ashore, and they were subsisting on fish caught under the ice. There was another camp a couple of miles further west. These Eskimo, by the bye, are more civilised than the Nechilli. Hospitality is a leading feature with them. When visiting them they always offered us tea and fresh wheaten bread. The Eskimo here on the Alaska coast certainly are bigger tea drinkers than any other people. We kept mostly on the ice, near land, where the route was smoothest and best. As we were not in training, we made a halt the first day at Key Point, fifteen miles from King Point and twenty miles from Herschel. We put up our tent at the Point, collected some wood, and made ourselves comfortable. I had brought Manni along with me, to show him the big vessels and the many Kablunas. The Eskimo, Neiu, also accompanied us.

Next morning we continued our journey. The sledging on the ice here was abominable, there being half-a-foot of water under the snow. We were wading in slush, and our reindeer-skin boots were soaked through. But we managed to get along, and at half-past four in the afternoon we trooped into Herschel Harbour. It was quite an unusual sight that met my eyes here. Five large vessels were lying side by side, and there was a multitude of people on the ice between them. Our arrival attracted considerable attention. Sten and Neiu, of course, were already known to most of them, but Manni and I in Nechilli costume were something of a novelty. We were quickly surrounded by a motley crowd—mulattos, negroes, yellow, and white men; their clothes were also of a very miscellaneous description. Most of the Eskimo were dressed as Kablunas, and most of the Kablunas as Eskimo. By the word "Kabluna," the Eskimo here mean all people of a strange race. When referring to negroes, however, they add "maktok," and describe them as "maktokkabluna," which means the "black white."

I had made up my mind to put up with Captain Tilton, on board the "Alexander," and there I went. I was received with the greatest cordiality, and conducted into a snug, homely cabin. Captain Tilton was a tall, powerful man, who looked older than he actually was, with sparse hair and a white moustache. The other captains soon appeared on board. All of them bore the usual traces of the conditions of life obtaining in these regions. They were corpulent, and their hair thin. That things on board some of these American whalers were not as they ought to be there can be but little doubt; but, having no positive proofs, I prefer not to mention the many and queer tales I heard during my sojourn here. I was, however, from the first start treated with the greatest courtesy by Captain Tilton. He was ready to assist me in every way, although he was not too well provided himself. My visit to the whalers was pleasant throughout. The receipt of a letter from home was the greatest pleasure I had. It was an old one to be sure, written nearly a year and a half before, but none the less welcome on that account. I also had a letter from Consul Henry Lund, in San Francisco; and the cap-

WINTER LIFE AT HERSCHEL ISLAND, 1905–1906.

tains told me how much this gentleman had done for me. They had, all of them, instructions from their owners to assist me in every possible way.

Before leaving Herschel I paid a visit to the local missionary, Mr. Whittaker. He lived on shore, with his wife and two daughters, in a house which, besides providing accommodation for himself, also had a schoolroom and a chapel for the Eskimo. I was present at one of the services, and it was a real pleasure to hear the Eskimo sing. As a practical man, which every missionary should be, Mr. Whittaker had studied his people and found that they were fond of singing; he therefore introduced as much singing as possible into his services and gave sermons that were short but to the point. The consequence was that the services were well attended. The missionary was a real English sportsman, tall, slim, and powerful, qualifications that might come in useful in these regions. It happened also that the minister had to act as policeman. The Eskimo have the nasty habit, as soon as they get hold of spirits, of immediately getting drunk, and then they are not easy to keep in order. However, Mr. Whittaker certainly held a most difficult post in a place composed of many doubtful elements which, in various ways, hamper the work of a minister and a missionary. I imagine that both he and his family were very glad when their time was up in the spring and they were able to go south. The two little girls were six years and nine years old respectively, and they were exceptionally pretty children. Both of them spoke

THEATRICAL PERFORMANCE AT HERSCHEL ISLAND.

English and Eskimo fluently. Unfortunately, the younger one was taken ill and died in the spring. It was a touching sight to see the parents carrying the little one with them on a sledge as they left the island.

Manni had amused himself royally. He had company all the time and had joined in the "hola-hola" or Eskimo dance. The Eskimo here, however, are so greatly influenced by the "Kablunas" that they lose much of their interest. Manni had been treated to the best of everything. He had had his fill of reindeer and seal, as well as of whale-blubber, and his fill was no trifle. The Eskimo lived in poor, low, little wooden huts on land. They did not look very healthy, and when the visit terminated, Manni was suffering from a bad cold.

The post was to leave *viâ* Fort McPherson for Fort Yukon on October 20th. It was to wait there and bring back telegrams to the different captains. There was not much probability of our getting the return mail before the month of May, when it reaches Fort McPherson *viâ* Edmonton and is forwarded thence by Indians to Herschel Island. This seemed to me too long to wait, and I therefore asked the whalers if they had any objection to my accompanying the post on October 20th. If I got down there myself, I thought I could arrange matters. My wish was readily complied with, and everything I required was placed at my disposal. Captain William Mogg, the commander of the stranded vessel "Bonanza," was also going to accompany the post and try to reach San Francisco, intending to return next year by another ship. It was really a venturesome plan for such an elderly man, who, moreover, was not accustomed to such trips.

On September 29th, at 9 A.M., we started on our return journey to the "Gjöa." Mogg followed, to get some of his provisions from King Point. Without any great effort we kept going all day, and arrived there at 11 P.M. The northern lights were magnificent during the night. We called upon our neighbours at the nearest Eskimo camp, and had tea and wheaten bread. We here encountered a sledge expedition, sent out by Captain McKenna to Herschel Island. We learned that the "Charles Hanson" had got stuck in the ice off Toker Point, on the far side of the entrance to the Mackenzie River. As to the other vessels, the sledge party knew that four of them had reached Bailey Island, but they knew nothing about the fifth one. The wildest rumours and stories were in circulation all the winter about the disappearance of this ship. It was a small schooner, no bigger than the "Gjöa," called the "Olga." The last time she was seen she was going northwards, and there was much anxiety lest she might have been crushed in the ice or have drifted towards the Pole.

On reaching home we had covered thirty-six miles, which, unaccustomed as we were to walking, was a pretty stiff task.

Good work had been done during my absence both on board and ashore. Lund had arranged a new meteorological box, constructing it so that no fog could penetrate

ESKIMO TENT AT KING POINT.

it. Next day, with Ristvedt's assistance, I fixed up all the meteorological instruments, so that we could begin our observations on October 1st. The building of our observatory for the magnetic variation instruments was making good progress, and was completed by October 2nd. They had been obliged to blast away four to five feet of the frozen soil, and the whole observatory was thus subterranean with the exception of the roof, which was visible. Boards from the "Bonanza's" cabin were used as materials for this house, the whole of it being covered with roofing felt and old sails.

Eskimo were now hurrying past every day in both directions. It was not all pleasure to have these strangers constantly coming and going. One day, for instance, no less than four families arrived simultaneously and fixed up their tents close to the "Bonanza." They began by letting loose the dogs, and as these were by no means overfed, they at once started, as might be expected, in search of provender. Captain Mogg had packed his sledge in readiness to leave next morning. Among other provisions he had some frozen fish, with which he intended to treat his fellow-skippers, as they had not yet had any. The dogs, as a matter of course, broke through the packing of the sledge and gobbled up the fish. Sten's larder was also broken into during the night by the thieves, and, according to his statement, they had stolen 250 frozen fish. In our wood shed they found a pair of brand new sealskin trousers, belonging to Lund, and these they had devoured. There was a great commotion next morning on the discovery of the loss. I was hiding behind a pile of drift wood watching the fat little Captain Mogg, when he came down to have a look at his sledge. He dived his head into the empty fish-box, swearing like an American trooper. Sten, with his carbine

A COFFEE PARTY AT KING POINT.

shouldered, was parading up and down in front of his house like a sentinel, but the Eskimo in their tents were laughing and singing, and preparing to depart, ignorant to all appearance of the occurrence. Sten was clearly waiting for them. I walked slowly up to him, assuming the most jovial expression I could command. "Well, Sten, good morning. How goes it?" Sten turned round angrily: "The d——d swine have stolen 25c. . . . But not a single Eskimo shall get away from here before making restitution." Sten hailed from the neighbourhood of Sandefjord. However, the Eskimo continued their preparations quietly and at their ease. When I came out again after breakfast, they were already a long way off, on the ice. I then asked Sten if he had received payment or restitution for his fish. No, not exactly, but every one of the Eskimo had guaranteed to bring fish back in the spring. He did not go further into details as regards the nature of their guarantee, but I suspect it was not worth very much. The incident was regrettable because Sten was always ready to be of assistance to anyone. When he heard of my decision to accompany the post, he at once started to make me a new sledge. I had already plenty of sledges, and good ones too, but I did not like to refuse his kind offer. As a sledge would hardly be much used on my post journey, it did not matter very much how it was built. His wife, Kataksina, sewed new boots and clothes for my journey, and I think he employed other Eskimo also for the occasion.

As it was a matter of importance to secure sufficient supplies of fresh food for the winter, I decided to send out our Nimrods, Hansen and Ristvedt, on a hunting expedition. The Eskimo, Neiu, had, presumably, had enough of the heat from Sten's kitchener, and accepted with pleasure an appointment as guide to the expedition. They were fitted out for two months, taking with them two of ours and four of Sten's dogs. I myself kept four for my journey. Our three lady dogs had, unfortunately, been heedless of time, and were daily expecting some interesting events, so we could not make use of them on this occasion.

During the first half of October we did unusually well at fishing. Our regular catch was thirty to forty a-day. We cleaned the fish as soon as we lifted it out of the net and hung it on the rigging, where it froze immediately. Everything considered, the "Gjöa's" rigging was a pleasing sight at this moment. It would have gladdened the heart of many a fishmonger and poulterer to have seen the splendid assortment of grouse, ducks, geese, and fish dangling there. We styled it: "Jensen's in the Torvgaten."

Manni had now become a pupil of Lieutenant Hansen. He was learning to write and to tell the time. He did not show any special talent, but he learned to write his name pretty quickly, and to tell the time to within five minutes, anyway. He made quicker progress at draughts. He acquired its secrets so well that his master and teacher had to submit to be beaten at times. While these games were in progress I was, as a rule, sitting in the cabin, reading, to the pleasing accompaniment of the subdued and well-considered observations to which the play gave rise. But when by some chance Manni won then there was a roar which for a while rendered impossible any attempt at literary enjoyment in his immediate vicinity.

However, previous to my departure, everyone was busy letter-writing. Each, of course, had one to write to his people at home. All the letters were finally put into a brass case and soldered up. You want a very substantial envelope up here. On the evening of October 20th my whole outfit was ready. The sledge, constructed by Sten, was shining with varnish and metal mountings. The perfectly new-sewn tarpaulin, made by his wife, added to the appearance of luxury. The weight on the sleigh was only 2 cwt. now. It would, undoubtedly, be heavier when leaving Herschel. I took Manni with me to let him have his first glimpse of civilisation. Besides, he was a good helpmate to take when travelling in these regions.

On October 21st, at 6 A.M., we said good-bye to our comrades, and departed. It was a lovely day. The road was excellent, as the cold weather had not thoroughly set in as yet. We kept along the shore, which was nearly bared by the wind, and proceeded at a great pace. Both of us were attired in Nechilli costumes, but it was not long ere we had to divest ourselves of our outer clothing. When we had trotted nearly as far as Key Point, we noticed something strange, shooting up from the inner side of the bay, right across our course. It looked very much like a balloon being dragged along on the ice. As it approached we saw it was a sledge, carrying sail. The sail fetched the sledge along in the fresh breeze at such a speed that our dogs had to be careful to avoid being run over. The sledge caught us up and followed our course. We tried to keep up with them, but dropped farther and farther behind. There were four Eskimo on the sledge. Suddenly they stopped, and the driver turned to me and proposed that we should tie our sledges together and harness all the dogs to them. I willingly accepted his offer, and the coupling-up was arranged in a few minutes. The two sledges, when combined, sailed away before a strong fair wind, towards Herschel Island. The amiable Eskimo driver ran alongside our sledge, and we conversed together. He was an exceedingly agreeable fellow. He spoke English quite well. He was of the type that inspire confidence, and reminded me of my excellent friend, the "Owl," at Ogchoktu. My pleasure was therefore equal to my surprise when Jimmy, that was his name, informed me that he was going to take the post from Herschel Island to Fort Yukon. In answer to my questions as to how long a time the journey would take, and what the road was like, he frankly told me he did not know, as he had never done the trip before. He had received instructions from the white men to take the post and was going to do so. I was informed afterwards that none of the other Eskimo had ventured to undertake this journey. As a reward for his services, Jimmy had been promised a whaling-boat, which, to an Eskimo, represents the summit of his ambition. With the acquisition of a whaling-boat they consider themselves fully set up for the remainder of their lives.

At 3 P.M. we reached Herschel Island.

Things went on quietly and regularly after my departure. Lieutenant Hanson took over the command during my absence. A number of hunting expeditions were sent out, and they never returned empty-handed. Kunak, who lived next door to Sten, was sent off to the mouth of Mackenzie River to hunt for elk, which are very abundant there. He continually sent elk-meat home, and Sten shared it with us. Other Eskimo also called and sold us elk-meat and a large quantity of hares.

The weather was unusually severe this winter. Christmas came, and by the united efforts of the colony it was celebrated with mirth and festivity. Snow and storm, however, continued their game without cessation in the new year and rendered our existence as unpleasant as only such visitations could. Everything is enveloped in an impenetrable haze; the snow is blinding and it penetrates into every chink and crevice. One night Manni had to go from our house to the ship. It was a rough night, one of the worst, and as Manni did not return when expected the Lieutenant became anxious about him. He went ashore to look for him and found him in Sten's house. He had lost his way, even at this short distance.

The mail arrived at the beginning of March. There were some newspapers and also a despatch from me sent through the Royal North West Mounted Police, who left Dawson City on December 25th. From these papers and from my communication they received full particulars of events at home, and moreover all had news from their friends.

On March 12th at 6 P.M. I was back on board and brought newspapers and letters for all. I found everything in perfect order. Lieutenant Hansen had during my absence taken the meteorological observations and Wiik had kept up the magnetic work without interruption. Before leaving I had given instructions to deliver ten cases, or 24 cwt. of flour to the whalers, as they were badly provided with this commodity. It was a matter of congratulation to us that the "Gjöa" could offer the Americans this assistance after two and-a-half years' sojourn in the ice.

The day following my return was observed as a holiday. There was much to be thankful for. Shortly after this Ristvedt went, with Jimmy, to Herschel Island to consult a doctor. He had got a grain of sand in one of his eyes and was unable to get it out again. On March 16th we took stock of our provisions and ascertained that we had plenty. Ristvedt returned a couple of days afterwards, relieved of his complaint and much impressed by the hospitality shown him by Captain McGregor of the "Karluk," and his wife.

We availed ourselves of the first fine weather to build ourselves a large airy house on the top of the hill, as a depôt for our collections. When the sun commenced to shine in real earnest, we were going to put out our skins and plants and give them an airing. We were quite as anxious about these collections as we were about our own lives. On March 22nd we had a maximum temperature of 32·36° Fahr.—just above freezing point—for the first time this winter. Spring began to make itself felt.

Neiu had just come home with ninety hares. I decided to let a couple of our huntsmen accompany him and his wife on their next hare-hunting expedition. We were not short of shot guns, the whalers having provided us with a supply. The day after their departure some Eskimo arrived, and sold us 150 lbs. of elk's meat and forty hares. In other words, we had abundance of fresh meat.

Wiik had not been very well the last few days, though it did not appear to be anything at all serious. He complained of a poor appetite. On the 26th he had acute pains in his right side. I gave him some medicine, which relieved him. The next day he had to stay in bed, the pains having returned. I presumed it was a touch of pleurisy, as, from what he told me, he had suffered from that before. I started treating him with cooling bandages, according to "Uckermann's Medical Guide." During the night of the 28th he had a good sleep, and by the morning was in the best of spirits: laughing and joking constantly. However, the pains recurred in the afternoon. I then took off his cooling bandages and put on a mustard plaster. His pulse at night was steady at 104° and his temperature registered 103°. I was called next morning at 4 o'clock. We had an electric bell between the house and the ship, and I had given instructions to be called should any change occur.

Chapter XI.

Wiik now complained of more intense pains. His right side was slightly swollen. The mustard plaster had produced no effect, probably it was too old. I then applied a mustard bag, and this at once commenced to act. His temperature remained at 103°. Later in the day I thought I saw a considerable improvement in our patient. He slept for some time, now and again, and was not in pain. I gave him some fever medicine which did him good. By noon the temperature was 102·5°. He asked for some food and ate well, under the circumstances. At 9 P.M. his temperature was 101°, pulse 116, but steady. I took off the mustard bag; it had acted very well, and I opened the blisters, dressing them with lint and boracic vaseline.

On March 30th I made the following entry in my log-book: "Wiik is making good progress towards recovery. Temperature this morning 102° with regular pulse of 116. Temperature this evening just below 100°, pulse 114, regular. Appetite improving. Evacuation all right." I was in really good spirits that night at the prospect of a speedy recovery. I was not called up in the night and was, therefore, certain of finding our patient practically well again; when I went ashore for breakfast next morning, I was greviously disappointed. He had an acute attack of shivering in the night. Lindström, who was lying next to him, at once covered him over with a lot of clothes. As this had no effect, Wiik asked Lindström to lie on him. Lindström did so, and the attack began to subside. Lindström then got the fire lit in the kitchener and thus raised the temperature in the room. He did not ring for me, as a tremendous snow-storm was raging at the time. Lindström, of course, did

GUSTAV WIIK (WINTER, 1905).

wrong, but he acted with the best intentions. Probably I should have been unable to do anything. When I arrived on the scene in the morning his temperature was 102°, pulse 116. When I took the temperature again at

11 A.M. it was down to 101·5°, but the pulse alarmed me; instead of being steady as before it had now become very irregular. I then got Jimmy, who had accompanied me to Alaska, to prepare himself for a journey to Herschel Island to fetch a doctor. A heavy snowstorm was blowing from the north-west, and Jimmy insisted on waiting till 2 A.M., as it was too late in the day to get there before dark. I wrote a letter to Captain Tilton, on whose vessel the doctor resided, and also one to the doctor himself, explaining how matters stood. Wiik's condition meanwhile became more critical; I used all my power of persuasion to get Jimmy to start at once, but the weather was so bad that not even an Eskimo would venture outdoors.

At about 5 P.M. when making the final preparations for Jimmy's journey next morning, the bell rang very loudly, all of a sudden. There was something in the ring that spelt disaster, and I ran off at full speed. But I was only just in time to see our dear friend breathe his last. It was an inexpressibly sad moment. I closed the eyes of our late comrade, and we remained sitting there for a while in silence and sorrow.

Wiik was everybody's friend. His humour and jocularity had afforded us many happy hours. Death must always be a gruesome guest, but to us, in our position, far away from friends and relations, it was, if possible, more depressing than it would otherwise have been. As soon as possible we resumed work, the great consoler and helper.

As I was not quite sure of the nature of the malady I thought it wise to remove everything from the house. Sten offered us accommodation in his house, and I gratefully accepted the kind offer. We made our kitchen arrangements in his house and had all our meals there. Lindström and Lund moved into our cabin, aft, until they could get things arranged in the deck-house, forward. Our house was thus entirely deserted, and we nailed it up.

Sten went to Herschel Island a couple of days afterwards. When he came back he reported that it would have been useless to send there for the doctor, as at the time he was overwhelmed with work. A couple of poisoning cases in particular had claimed his personal attention night and day.

Lund had the black-painted coffin ready on April 3rd, and Wiik's body was placed in it. We deposited it on two stools in the outer room, screwed on the lid, and covered it with a flag. We had to keep the coffin until the sun had softened the frozen ground sufficiently to enable us to bury him. All chimney-pipes and other small openings were left uncovered, for the sake of ventilation, and the house nailed up again.

Our two sportsmen returned on April 5th. They had heard of the death of our comrade from some Eskimo. They brought with them 237 hares on their sleighs. They reported that there was an enormous quantity of them. The huntsmen caught them by forming themselves into a chain right across a long strip of underwood and drove the hares before them with shouts and yells. The stupid animals would not break cover and finally got all huddled up together and fell an easy prey. These hares are not much smaller than ours. We reckoned one hare for every two men.

The first thing the whalers do when preparing to winter up here is to build a cold store. It may sound somewhat superfluous to have a refrigerator in the Arctic Sea, but it comes in very useful in the summer. We had, however, omitted to build one in time, so we had to look about for a storeroom for all the fresh meat we had on hand. We seized on the plan which we had found to answer before, we went on board the "Bonanza" to see if there was anything suitable. We were not disappointed this time either. There was a capital, spacious cellar below the cabin floor. During the autumn it had filled with water and was now frozen. In other words, we had the most magnificent cold-storage cellar, all complete and ready for use. We bundled up the hares and hung them there; so we had fresh meat whenever we wanted it.

The first messenger of spring was a raven, which arrived on the scene on April 4th.

Lindström very soon moved up to his friend Sten. This was a practical step, inasmuch as most of his work had to be done there. A firm friendship had sprung up between the two stoutest fellows in the colony. When we went aboard after supper it rarely happened that any of us returned ashore until the next morning. Lindström and Sten were then in undisputed possession of the house.

One night I had forgotten my pipe and I went back to fetch it. As I entered the wood-shed I heard roars of laughter inside. Ahem, I thought, perhaps I ought to have looked into this before; what is it that makes these two so inseparable up here? Regular orgies, evidently. "Can-can" with the girls . . . . I suddenly appeared before them like a God of Thunder, ready to punish the guilty, and there they were, playing a game of cards, as happy as a couple of schoolboys. I enjoyed the spectacle for a time, and then left with my pipe and my own reflections.

The hunters returned on April 12th, this time with seventy-one hares and five ptarmigan. Manni was an excellent fellow. He was always happy, careful, and kind. What he loved was hunting, and he was nearly always in the field. As he was going with Ristvedt and Hansen this time, I gave him strict instructions to give up chewing tobacco. I never liked this nasty habit, and it seemed very undesirable that the boy should acquire it. Ristvedt and Hansen both chewed tobacco just as others would eat bread and butter, especially when out hunting. They tried to tempt Manni, but he would not touch their tobacco. I had given him permission to smoke, but I told him also that too much of it might be dangerous. And he was very moderate. As I was always afraid of his catching cold, I told him to change his clothes when he came home. Ristvedt reported that Manni always changed everything on his return home. Notwithstanding this, he had got a thorough cold now; he had probably caught it from the Eskimo, who don't take the slightest care of themselves.

Easter came and went without any special celebrations on our part. A cigar each on Easter Day was the principal luxury indulged in. Mr. Howard, Inspector of the Royal North-West Mounted Police, passed us on

SPRING AT KING POINT.

April 14th. He was accompanied by a sergeant, an Indian, and an Eskimo, and he was in uniform; it looked very peculiar up here. He came from Fort McPherson and was on his way to Herschel Island as a representative of the Canadian authorities. A sleigh also arrived from Herschel Island with some things belonging to Mr. Whittaker, the missionary.

It was going fifteen miles further east, to Shingle Point.

Spring had commenced, and with that the traffic. On the following day a sleigh arrived from the east. The party consisted of some Eskimo and a white man sent by McKenna; they were going to procure some sugar for him. The white man had rather a crazy look about him. He had had one of his big toes rather badly frostbitten, and was treated by Lieutenant Hansen. He told Sten subsequently that he had been in the company of the second engineer of the "Charles Hansen," but that the latter was taken ill on the road, so he had been obliged to leave him behind after covering him over with a blanket. This information seemed rather scant. Charley, as the fellow called himself, had nothing on his feet excepting a pair of ordinary stockings and sealskin kamiks. This was very little in these regions. When he reached Herschel Island it turned out that he had deserted from McKenna together with the second engineer, who apparently had died somewhere on the shore. His meeting with the Eskimo, who were going to fetch the sugar, was purely accidental.

Lieutenant Hansen and Ristvedt went off to Herschel about the middle of the month to see if we could secure two more men. I wanted to have another man in the caboose, as Lindström prefers to look after the engine, and also another man on deck.

Meanwhile, rumours had reached us that reindeer had

been seen in the neighbourhood. I, therefore, told off Hansen, with a family of Eskimo, to go hunting. All our dogs, with the exception of Nicodemus, which I had from Eagle City, were in a miserable condition, partly owing to fighting and partly to other misfortunes. Hansen's expedition was, therefore, dubbed the "Invalid Corps." As a matter of fact, he returned the very first night, and reported that the petroleum had leaked out and spoilt all his bread. The only thing we could do was to give the "Invalids" a fresh supply of bread and let them make another start next day.

Mr. Whittaker arrived on April 22nd with his wife and daughter. They remained with us the night and proceeded next morning. Ptarmigan now made their appearance in big crowds; the hills seemed alive with them, they were so plentiful, but they were so shy that it was almost impossible to get within shooting range. Manni somehow managed to bring home as many as four brace in a day. This lad made progress in his studies every day, but never acquired much of the white men's language, His greatest advancement was in the games of draughts and patience. It was often difficult to find anything to keep him occupied, when he had finished his regular day's work of cleaning, wood chopping, water carrying, and hunting. It generally ended in a game of draughts with the Lieutenant or a single-handed patience.

Lieutenant Hansen and Ristvedt soon returned from Herschel Island, where they had been handsomely treated as the guests of Captain McGregor of the "Karluk." The American whalers with their customary kindness consented to let me have two men, whom I might fetch on June 1st. With these two new arrivals our accommodation on board would be inadequate; we should have to increase it. A word about this to Lund was enough; he had his plan ready at once. We agreed

A SUMMER SCENE AT KING POINT.

upon building a small room on either side of the cabin stairs, one for Hansen and one for Lund. The few boards we had brought with us from Christiania, already used for the observatory on King William Land and in our residence here at King Point, were again to be pressed into the service. There was just sufficient for these two small berths. The rooms were not large, nor were the fittings luxurious, but when ready, decorated and painted, they looked really nice and cosy.

The snow decreased noticeably every day as we neared the end of April. There were large numbers of "hicksies," a kind of earth rat, on all the bare spots. They were very fat, and looked as if they had done nothing but feed all the winter. Their skin is used a good deal for coat linings, and the animal is therefore much sought after. It takes at least sixty skins to line a coat.

Manni's hunting during the spring consisted chiefly in catching "hicksies." Some he shot and some he caught in snares. He placed snares outside their holes and lay down at a distance, holding a string. When the rat looked out to see what was the matter Manni pulled the string and the little creature had the noose round its neck. The weather kept warm. The temperature remained above freezing point. Pools of water commenced to form on the ice.

After about eight days' hunting Hansen returned with fourteen reindeer. The Eskimo, Anakto, had brought most of these down with one of our Krag-Jörgensen carbines. The Krag-Jörgensen guns have made a good all-round reputation for themselves this winter, far better than the Winchesters. Hansen had enough to do to get on board all the meat the Eskimo procured for us, so he could not take part in the chase. A large number of Eskimo were now hunting on behalf of the whalers, and to compete successfully on this

ANAKTO. AN ESKIMO FROM HERSCHEL ISLAND.

difficult ground we needed to be Eskimo ourselves. The reindeer were very shy and timid beyond measure.

I was very much surprised one day at the end of April to receive a visit from a man I had met and shaken hands with near Rampart House, on the Porcupine River, away in Alaska, when I was on what I called my post-journey. His name was Mr. Darrell. He was a most remarkable man, possessed of rare vigour, courage, and perseverance. He may have been about forty, short but powerfully built, and very fair. When any of the American whalers, who wintered off Bailey Island, wanted to send their post southwards, they had it sent along to Fort McPherson by Eskimo, with a request to the commander there to be good enough to forward it on to Fort Yukon by Indians. Owing, however, to the heavy fall of snow in the course of the winter none of the Indians dared to undertake the journey across the mountains between Peel River, a tributary of the Mackenzie River, and Porcupine River. Mr. Darrell was at that time employed by the Hudson Bay Company. Hearing of the Indians' refusal and being fully aware of the importance to the whalers of the interests involved, he decided to take the post himself. He fitted himself out with a toboggan and dogs, and started off without a single companion. It may seem a foolhardy undertaking, but Darrell had journeyed by sledge before, and probably had his own reasons for preferring to travel by himself rather than in company, which at times might be somewhat doubtful. The mountain ranges between Peel River and Porcupine River are, as a rule, no worse to cross than most others. But the Indians were right, the enormous snowfall had created serious difficulties. Mr. Darrell soon came to the conclusion that his dogs were useless. They simply scraped and rolled about in the snow. He promptly decided to leave them, and went on alone, dragging the toboggan behind him. The toboggan, of course, was small, and the outfit as light as possible, yet it must have been hard work, but he got through. He rested a couple of days at Rampart House, a small trading station on the Porcupine River, and laid in fresh supplies. It was a couple of days' journey from there that I met him with the post on my way to Eagle City. He came jogging along alone with his toboggan, and was due to reach Fort Yukon in a week's time. We did not have a long conversation that time, but he told me he would return *viâ* Herschel Island, and I invited him to stay with us as long as he remained. He thanked me, and I never thought I should see him again, yet, on April 29th, he came quietly marching up with his toboggan exactly as I had left him on Porcupine River. He made us happy by remaining with us for a couple of days, and then left quietly and unassumingly to continue his journey as before. I stood looking after him as he disappeared from view, and I thought, if you got together a few more men of his stamp, you could get to the moon. Shortly afterwards I received a letter from him, on a small slip of paper, brought me by an Eskimo from the fort. It did not say much. He thanked me, and mentioned but briefly that he was near losing his life on the latter part of his journey. The Eskimo who brought me the letter told me a few circumstances which the unassuming man had not mentioned. He had lost his way at the entrance to Mackenzie River—anyone not thoroughly acquainted with the locality would easily mistake the route—and he was saved at the last moment by some Eskimo.

On the morning of May 2nd I was awakened by somebody falling down the cabin stairs. This had often happened, as the staircase was steep, and I merely opened my eyes to see which of us it was this time. There on the floor was an Indian giving vent to a flow of language I quite failed to appreciate. When he had finished I quietly asked him in English what brought him there. He answered in very good English that he had brought the mail for me. I was soon wide-awake and got hold of the two letters he had brought. This was the first regular mail that year for Herschel Island *viâ* Edmonton and Fort McPherson. Lieutenant Hansen and I were the fortunate ones ; we had a letter each. Mine was from my brother. It was a bit old, but not less acceptable on that account. If people could only picture to themselves how much letters are prized in such circumstances, I fancy the number of letters written would be much greater.

Hansen laid in his load of meat regularly once a week. Eventually we had accumulated between 1,500 and

1,600 lbs. of reindeer-meat, so we were well supplied. The Eskimo sold us the meat at 5 cents a pound, which was the current price locally, and they also had their own supplies.

The mail-carriers returned from Herschel Island on May 6th and were to proceed direct to the Fort. We gave them food and plenty of provisions for their journey. They took away with them a couple of letters and a despatch I wanted sent off as quickly as possible. It was my notification of poor Wiik's death. I was very anxious lest his mother might hear news through some other source that everything was going well with the Expedition, before she received the sad news from me, as that would make the blow all the harder and the disappointment all the bitterer, when she learned the truth. I left the telegram open and sent it to Mr. Firth, the manager of Fort McPherson, accompanied by a letter in which I explained the circumstances and my reasons for wishing to have the telegram dispatched at the very earliest opportunity. I asked him to advance the necessary outlay for the same. The telegram never reached its destination. The mail from the Fort arrived at Herschel, in August, but there was not a word for me from Mr. Firth.

The month of May was lovely. As soon as spring weather set in in earnest we got all our things out to air and dry. This was very necessary, as a dirty snowy winter like the one just ended makes everything damp. Lindström stretched some fishing nets over stools and spread our collections out to dry. Even the empty egg-shells had a thorough airing.

HELMER HANSEN (SPRING, 1905).

The two new men engaged from the whalers arrived on May 8th. I was certainly somewhat astonished as I did not expect them before June 1st. They had their discharges in proper order. One of them was a Norwegian, Ole Foss, hailing from Fredriksstad. He made a very favourable impression and showed himself, during the whole of his service on board the "Gjöa," to be a capable, reliable and decent fellow. The other one was a young American, named Beauvais. He was to relieve Lindström in the galley.

Dr. Wight had written to ask me if he might come south with the "Gjöa." He had received news of illness in his family, and wanted to get home as soon as possible. As it was most likely that the "Gjöa" would get back to civilisation before any of the whalers, I wrote to the doctor and told him he was welcome. Now, however, we should be quite full up.

I had decided to convert the magnetic observatory into a mausoleum for Wiik. It was suitable for the purpose in every respect. Wiik had built it and used it himself, and was very fond of it. It was situate on the best and most open position facing the Arctic Sea. By May 8th we had completed our work in the frozen ground.

Next day, May 9th, at half-past ten in the morning, we assembled for the funeral. Every flag was flying at half-mast. We carried the coffin out of the house, and secured it with ropes on one of our sleighs. We then drew it as far as the mausoleum. Once again our comrade was on his way from the house to the observatory, but this time he would never return. We halted outside the entrance, while I said my farewell to Wiik and read the Lord's Prayer. The ceremony was brief, but I think it will long be remembered by us all. The coffin was brought in, placed on two small wooden stools, and covered with a Norwegian flag. The room was then filled with drift-wood, and walled up. Later in the summer we erected a high cross on the northern side of the grave, put turf on it, and covered it with flowers. The American whalers promised me to tend it every year and keep it in order.

WIIK'S GRAVE AT KING POINT.

The hills were now getting green, and the brooks began to trickle and murmur. The water from the brooks tastes appreciably better than the ice-water, and the water we took from the sea was sometimes so brackish that we had to give up using it. No drinking water is equal to that which trickles fresh and clear through the soil.

Chapter XI.

It sometimes happens that seemingly trivial yet very interesting observations are made quite unconsciously. One day, when the thermometer stood at 21·2° below freezing-point, I put a liqueur-glass filled with water on the ship's rail. I was in the habit of taking a glass with me to the observation-box to moisten one of my thermometers. The rail was painted green, and I was surprised to notice that the water in the glass did not freeze, notwithstanding the sharp frost we then had. When I removed the glass, and placed it on a white surface, the water at once commenced to freeze. The sky was cloudy all the time.

On getting our residence vacant again we fitted up our bath there and used it regularly. I had the first bath; then it was Manni's turn. The others had made him believe he was to be boiled. He therefore went to the bath with considerable misgivings. When he found that he had been made a fool of, he laughed heartily.

Neiu, the Eskimo, had come to loggerheads with Sten over a bag of flour. Being offended, he cleared out of the house and set up his tent on the shore among the driftwood. He hunted from there, and one day he brought home a lynx he had shot some way out on the ice. We concluded that the lynx must have had a fit of insanity.

May 17th was celebrated as usual as a gala day, with flags and a banquet.

There was a good deal to do, however, this spring. The change from petroleum to wood necessitated several

MANICHYA AND FAMILY AT KING POINT.

alterations: we had to alter the galley. A petroleum tank was fixed up as a kitchener and provided with a chimney pipe which bent and twisted out into the air with the most fantastic contortions. This work of art was conceived in a most ingenious spirit, the intention being to secure a good draught in whatever quarter the wind came from. It was excellent in theory, certainly, but in practice, this chimney with its cowl, had an annoying yet unfailing tendency to collide with the mainsail. Any yachtsman will appreciate this.

The hold had to be completely re-stowed, and all our collections, etc., had to be brought on board. The Lieutenant continued reading most diligently with Manni, and I could not help admiring his patience. I can still hear them: a, b, = ab; b, a, = ba, abbaba. After half a-year's assiduous work they were still at "abbaba," and Manni would even now make mistakes in this difficult word. I had lately fancied that Manni feels inclined to remain here with the Eskimo. I did not like to lose him, but on the other hand I was averse to keep him against his will. So I asked him one day if he would rather join the Eskimo again, and Manni said "Yes." The following day I went to see an Eskimo named "Manichya," a very capable man, and asked him if he was prepared to take Manni. Manichya was delighted. He had only one child, a daughter, and an addition to the family like Manni was not to be despised. Manni left the "Gjöa" the same day, equipped with clothes, tobacco, matches, soap, two guns and ammunition. He jumped and danced with joy. But I could not help thinking that the good fellow's life in future would be different to what it had been with us. Work from morning till night and perhaps little to eat when he came home hungry. The first thing his new parents did was to cut his heavy, magnificent hair. He looked pitiful after this ordeal. Next day he went away with them westwards to hunt seals. About three weeks afterwards he came and visited us. He brought me a few birds. He had "gone off" considerably already, and his bright eyes had a sad expression in them. His appetite was simply terrifying; he swallowed everything we put before him. When ready to leave he came round to bid each one of us good-bye. I could guess by his looks what he wanted, but I preferred that he should himself ask to be taken back. A fortnight later he returned. Again he brought a large bundle of birds for me. But his appearance was now deplorable. His cheeks were hollow, pale, and thin, and I there and then said "Would you like to come back to us again?" I shall never forget the glad and thankful smile he gave me as an answer, and thereupon the lost sheep returned to us. We had all of us become so fond of Manni that it was a matter of satisfaction all round to have him on board again. Even the Eskimo-hater, Lindström, smiled that day. Manichya was certainly a bit surprised to learn that Manni had again become a Kabluna, but he made no difficulties.

The only way now to keep our reindeer-meat was to

dry it. I got the Eskimo women to bone a large portion of our joints and hang them up, This dried meat was very useful.

Spring came much earlier here than at King William Land. All the birds of passage had arrived by May 20th. There was a peculiarity about the seals here. The males

"Gjöa." "Bonanza."
VIEW FROM THE TOP OF KING POINT. SUMMER.

smelt so unpleasantly that even the dogs would not touch their flesh. This was probably due to the pairing season; but, although these seals were of the same species as those at King William Land ("snadd"), we did not notice this smell in them there.

At the end of May the Lieutenant and I moved ashore, as the cabin was going to be painted. This painting was Lindström's first job as "handy man," and he did it very well. Beauvais took over the kitchen, It was very agreeable to be ashore and wake up in the glorious air, while the birds were singing beautifully. June was cool at first. The maximum temperature on the first of the month was 29·7°. The dogs had now finished their term of service, and I made Sten a present of them. The only one I kept was Silla, and her little son Ole. I had promised to take Nicodemus with me to San Francisco, so he, too, came with us.

On June 6th we moved back on board again into our snug cabin. Our Royal Family was in the best frame that King Point could produce, and the picture was hung on the middle of the wall, surrounded by a decoration of flags, with "Alt for Norge" (All for Norway) underneath it. It all looked exceedingly pretty. On one side of this picture there was a chart on which was marked the "Gjöa's" course through the North West Passage. On the other side was Nansen's likeness.

When the snow had melted off the ice along the shore, the ice itself soon began to disappear. It was all formed out of fresh water from Mackenzie River, mixed with a good deal of mud, so that it melted quickly. It very soon became quite porous and difficult to walk on.

Our best time was at an end, as gnats now made their appearance again. They arrived on the evening of the 28th, during a storm from the south-east. They got worse every day, and if we had not had some gauze material, which we used as mosquito nets, we should hardly have survived this plague.

I brought all the magnetic instruments on board on June 30th. Where the stand had been, I erected a wooden slab, marked "Gjöa, 1905—06." The

MARK, SHOWING POSITION OF MAGNETIC INSTRUMENT STAND (KING POINT.)

meteorological observations were brought to a close the same night.

On July 2nd we had a strong gale from the south with a temperature of 64·4°. We cast off from the ice and proceeded alongside of the "Bonanza." This old wreck had given us many a helping hand, and was going to do so again. When the land-lead increased we hauled astern of the "Bonanza," where the ice left us at peace; it was drifting to and fro on the tides. With everything on board we now drew seven feet

"Alexander." "Jeannette"
THE FIRST TWO WHALERS ARRIVING AT KING POINT.
(JULY 11TH, 1906.)

forward and 8 feet 10 inches aft. The main pack of ice was drifting out and in, and sometimes threatened to come right up to us near the shore. Fortunately it did not. We could see much open water beyond it, but there was nothing for us to do out there before the whalers turned up. If they could not get on, certainly we could not. There was a great movement of Eskimo during these days.

Finally, on the evening of July 10th, we sighted three whalers in the open water outside the ice. It was still doubtful whether they would succeed in squeezing through. But one of them continued eastwards along the ice and entered the channel near us about 5 o'clock next morning.

Our opportunity had now come, and everything was clear for a start.

CHAPTER XII.

WITH THE ESKIMO AND THE INDIANS. ON SKI AND SNOW SHOES THROUGH CANADA AND ALASKA.

WHEN I arrived at Herschel Island on October 21st, 1905, the preparations for the approaching mail trip were not yet completed. The captains had invited me to come a few days beforehand so that I might be there to see after the equipment and discuss the route with my companions. Again I went on board the "Alexander," where I was received with the usual hospitality.

Captain Mogg was to go south with me, and he had been requested by the other captains to see to the dispatch of the mail. In other words, he was the leader of the Expedition and I was to go as his guest. Captain Mogg was an old Polar traveller, and had also made several sledge trips inland, although it was certainly several years since he made his last. He showed the greatest enthusiasm, and worked hard with the preparations. I had brought several things from the "Gjöa," but Captain Mogg did not see the desirability of taking them with us. As his guest I could not enforce my wishes, so I decided not to make any comment but to leave myself to be entirely guided by the experience of the leader. There was, however, a tin of about 14 lbs. of pemmican, which I was rather annoyed

ROALD AMUNDSEN LEAVING EAGLE CITY. 1906.

at having to send back because Captain Mogg would not admit that pemmican was the best provision for sledge trips. Here, again, I gave in, as the Captain quite put his foot down on my good pemmican, and I considered it best not to begin the trip with any disagreement. I had enough pemmican and fish meal to last my five dogs quite a month, and Mogg had some dried fish for his seven dogs ; but it was obvious that he could not carry enough of this food with him. As, however, we calculated on reaching the Indians in three weeks we found that, by feeding the dogs all together, we should have enough, as by that means we could cut down the rations a little. Our own stores consisted of beans and pork, which were cooked together, frozen and packed in small portions, as also wheat biscuits, rice, sugar, butter, tea, coffee, chocolate, milk, figs, raisins, and spice. It was certainly a much richer list of stores than I was accustomed to, but I had my doubts as to whether in solidity this variety could compare with the simpler stores we used for our sledge trips. We also took with us a tent and tent poles, stove, lamp, sleeping bags, and many other luxuries. As we should not be able to use the sledges very long, for when we reached the loose deep snow in the mountains we should have to adopt the toboggan (the Canadian forest sledge), we lashed two of these to our load. The toboggan is made like a ski, twelve feet long, and six times as broad as an ordinary ski, but with a considerable curve. This was the first time I had seen this means of travelling, and I was anxious to see how it would turn out in practice.

On the evening of the 23rd everything was ready for

starting, and the extensive correspondence of the fleet of whalers was put on Mogg's sledge under lock and key. The Eskimo and Mogg had skin clothes like the local Eskimo; I was in Nechilli dress. Manni complained of pains in both legs after the long and tiresome trip from King Point to Herschel Island, and he begged to be allowed to return to the "Gjöa"; so not caring to take him against his wish, I obtained a seat for him on a sledge that was returning to King Point the next day.

The last evening we all assembled in the cabin of the "Alexander." Captain William Mogg, our leader, was a man with a big body, a little head, and small thin legs, and when he moved he gave one the impression that he was stumbling forward. He was English by birth, but had left home at an early age, and eventually became a whaler. During our stay on Herschel Island, Jimmy, the Eskimo, bore out the very favourable impression he had made on me when I met him for the first time on the ice. I had pictured Jimmy's wife to myself as a charming young creature, but Kappa might more easily have been taken for his mother than his wife. Kappa had come with a whaler from Kotzebue Sound to Herschel, and had met Jimmy, who was a Kagmallik. They were legally married both at Herschel Island and at Fort Yukon, so that it would have been difficult for them to get a divorce. Kappa, in her skin clothing covered over with calico, looked like a hay-rick. I took her to be over forty, but as a matter of fact, she was considerably younger. We got very friendly, and I regarded her as an elderly aunt.

JIMMY, WHO TOOK PART IN THE MAIL TRIP.

At 9 o'clock in the morning of October 24th we were ready for starting, and a number of men from the crews of the whalers turned up to see the mail go off. A breeze was blowing from the north-east, with

KAPPA, WHO ALSO SHARED IN THE MAIL TRIP.

a temperature of — 4° Fahr., but as we were going south-west, it did not trouble us very much. We went over the even snow covering of the ice at a dashing pace. In front of my sledge I had five dogs, all well broken in, and as it was considerably lighter than the other, I soon took the lead. The other sledge was drawn by seven dogs, but some of them were better than others, and they had not been broken in together, the consequence being that they gave Jimmy and Kappa plenty of trouble. Mogg was on my sledge. Dog-driving here is so arranged that the guide runs in front of the team and shows the way. We first followed the east side of the Island as far as the south-easterly corner ; here we crossed over the narrow sound on to the mainland. The snow had not yet covered the ground uniformly, so that grass tufts were often in the way. However, I strapped on my ski and had no further trouble. The others used snow-shoes. Here in Alaska the snow-shoes are narrower than the broad Canadian ones, and have a curved point, which is of great help in travelling. I never learnt to use the Canadian snow-shoes properly, but on these Alaska ones I got along very well.

We had first to mount a steep ridge, and not being in trim, it was rather hard work for us all. At last we reached the top, and the descent on the other side was, of course, much easier. At the foot of the slope we came on to the frozen bed of Herschel Island River. We had to follow the whole course of this river. Its delta here was a perfect chaos of sandbanks and gravel heaps, and it was very difficult to keep the right course, but both Jimmy and Mogg were well acquainted with the region, as they had been here on many a reindeer-hunting trip. This is one of the best hunting grounds, and thousands of slain animals are carted from here in the course of the year down to the sea to the whalers. This year the chase would be hotter than usual, as many badly provisioned ships are wintering here. Further towards the south, the bed of the river was clearer and more defined. In many spots the ice was clear, and now and then patches of ground stood up in our way, spoiling the runners of my sledge. Owing to lack of other material, we had had to cover the runners with galvanised iron, but this turned out even worse than I expected. It got quite worn through, and the projecting jagged pieces made it very hard travelling for the dogs. I endeavoured to remedy this by smoothing the runners with a stone, but as we were not very far from our halting-place, I deferred repairs and endeavoured to keep up as well as I could with the others, whose runners were in good condition. At half-past four in the afternoon we reached the mound in the river on which Jimmy had decided to make our first camping spot, as he knew that drift wood was always to be found there. The first day of a sledge run is always very tiring, and we were longing for a rest. In order that the necessary preparations might be effected rapidly, we so divided the work that Jimmy and Kappa undertook to erect the tent, assisted by Mogg, who took all that was necessary from the sledges, whilst I collected wood. On this particular evening it was very easy work, as small wood was lying about all over the bank, as if it had been cut for us; but collecting fuel at night would not always be quite so easy.

When I was ready, I helped the others to put up the tent, though it was done on a system quite unknown to me. Their tent consisted of the canvas and—eighteen tent poles. The place was first cleared of snow as far

TENT USED FOR THE MAIL TRIP.

as possible, the snow being banked round the tent like a wall. Then they inserted the poles, which were really thick willow twigs, and bent them so as to form an arch. Twelve of these poles were arranged opposite each other, lengthwise of the tent, and six crosswise. When they are all driven in firmly, they are bent over and lashed together, so that each pair forms a complete arch; then the canvas is stretched over, and the tent is ready, presenting the appearance of a hayrick, or igloo. This kind of tent has one advantage: the poles can be bent together higher or lower, according to the weather, and made capable of resisting any wind; but, in other respects, it is by no means a handy system for such a trip as ours. First of all it takes too much time to set up, and, secondly, all the necessary lashings must be done with the bare hands. The interior of the tent is never high enough to stand upright in—in fact, ours was never more than four and a half feet high—but we got accustomed to it in time; then there was the transport. Possibly a ready-made tent is more difficult to pack than this flat canvas, but even that is doubtful—but then the eighteen poles! They looked like bristles with bent tops, and when, after much trouble, they were finally packed away on the sledge, they made it look like a hedgehog, and were constantly getting caught in something or other on the way. Jimmy was a quiet man, but, whenever he had to pack these eighteen poles he swore, both in Eskimo and English, to his heart's content. An ordinary three-pole triangular tent is far preferable to this mushroom tent, and, while travelling by land, it is easy to find a valley or other shelter from the wind, thus doing away with the importance of the only ostensible advantage it possesses.

Jimmy and Kappa were, however, well accustomed to this kind of tent, and had it fixed up in a comparatively short time. Mogg took from the sledges what was necessary for supper and for the night, and Kappa put the tent in order whilst Jimmy made a fire. Mogg did the cooking that evening, and, in the meantime, we munched a few dried figs, of which we each had a handful, and they were very acceptable. When the dogs had been fed and the sledges carefully tied up for the night, we shook the snow off our clothes and went in. There was not much room, but I took the inside place, next the stove, close to one of the long walls, with Mogg next to me; Jimmy and Kappa lay along the other wall. I had to double my legs up; if Mogg wanted anything, he simply rolled over; and as to the Eskimo, long practice had made them contortionists. After the meal was finished, we all went peacefully to sleep.

I awoke the next morning at half-past four and looked round me, but none of my companions seemed to be troubling about the morning arrangements as yet; indeed, the uniform snoring of the trio indicated the very reverse. I took matters quietly and let the time pass on. Shortly afterwards Mogg awoke, looked at the clock and then at me, but I pretended to be in a profound sleep, so Mogg joined in the trio again. A quarter of an hour later the Eskimo awoke; they whispered a few words to each other and then turned over to sleep again. I calculated that the morning's work would occupy about two hours. If we were to get off at a reasonable time someone must begin, and as my companions were immovable, I turned out and set to work. I understood that the pleasant post of cook for the morning was intended for the guest, and I was allowed to retain the position the whole time. It was fortunate that the Expedition had someone who could get out of the sleeping bag in the morning. The morning work was not very complicated, and principally consisted in warming up what was left over from the previous night. Whilst I was occupied with this I had time both to think and write. The others were snoring so loudly that they shook the tent. When I was nearly ready I began to wake up my companions; this took some time, as the heat from the stove seemed to act like a narcotic on them; but at last they got up and we breakfasted; then we loaded our sledges, took the tent down, and drove on.

Several of the older ship's officers at Herschel Island had expressed their fear that we were beginning our trip too early; from the experience they had gained in many sledge trips, they did not think the rivers would be quite frozen over yet, and we soon found they were right. The river began to curve through sharp rocky passes and at many spots was open, so that the passages were exceptionally narrow. It was a delight to me to see real rocks again after the lapse of two years. The steep banks reached some four hundred feet high, and consisted of solid rock as opposed to the earth hills and moss heaps we had travelled over. I also knew that on this day we should reach the wooded district, and I was very excited at every turn in our course. When at length the first fir tree stood out against the sky up on the ridge—a very diminutive, battered little

Christmas tree, hanging out of a crevice—it produced a wonderful sensation, reminding me that we were now out of the Polar regions and on more homely human ground: at that moment I could have left everything that was in my charge and scrambled up the rock to catch hold of that crooked stem and draw in the scent of the fir trees and the woods.

Now and then in the narrow passes we encountered strong southerly gusts of wind which overturned dogs, sledges, and men on the slippery ice that offered no foothold. This was very tiring, and considerably delayed us. After running in front of the sledges all that day, a little rest was very agreeable in the evening, when we had found a good camping place on a little head of land looking on to a small pine wood. At first I had a good deal of trouble to collect wood. The Eskimo who constantly travel this route had so stripped the woods of dry timber that you had to search high and low to find enough for the night. I started with the axe on my shoulder, but was very conscious that my legs had already done good service that day, The snow lay deep between the trees and made very heavy walking; certainly I had my ski on, but I must admit that under conditions such as these, snow shoes have their advantages. They are easier to put on and take off, and one can twist and turn more quickly in such country as this, but at other times I would not have been without my ski on any consideration. The Eskimo here had often seen ski. When wintering at Herschel Island many of the crew passed the time running on ski in the hills. There were often Norsemen among them who could show the Eskimo some first-class ski running. For practical use, however, as for example, on such a trip as we were undertaking, they placed no reliance on ski. They often looked at them, and turned them round, but shook their heads at them, although before I separated from my companions they began to have some respect for my ski.

The next day we were stopped by water on the ice: it was not the open river, but water flowing on the surface. This inundation of the ice occurred frequently, even in the severest frost. We kept going till we were knee-deep in water, but at mid-day we had to give up and go on land and pitch our tent. We were in a narrow pass with high sides, and in the evening a splendid northern light spread its quivering strips of colour from one rock to the other. The next day the water had frozen, and with a little care we made some further progress. It was very wild scenery, large fissures showing in the rocks, filled with broken stones, large and small. The rocks extended right up to the river. They were not very high, but increased in height as we advanced. We followed the course of the river, the land rising so gradually as to be unnoticeable.

On the morning of the 27th we passed over a little side valley, running out westwards. Here the landscape suddenly appeared like a piece of genuine Norwegian scenery, timbered and rocky. The little valley was closely covered with trees, and from the lowest point there rose a huge snow cone to a height of quite 2,000 feet, while in the bosom of the valley nestled two little tents, like pictures from a fairy scene, with the smoke rising peacefully from the chimneys. Of course, we could not pass these simple folk without speaking to them, so we approached them; indeed, it might be they had fresh meat to sell us. We found the Eskimo occupied with their morning duties. As a rule, none of the Eskimo get up very early, they prefer to keep late nights. With their usual hospitality we were invited to partake of tea and fresh bread; the Eskimo make this very quickly: some flour, water, and baking powder are mixed in a pan and soon converted into excellent Polar cake; with a little syrup this is by no means bad. While chatting over the tea they told us that the river was open immediately ahead of us, and that we would have to pass over a ridge of land in order to get to the ice on the other side. They knew the country, and offered to guide us if we waited until the next day. We were easily persuaded to agree to this. These people consisted of four Eskimo, two men and two women. They were there hunting, and on the day before had had the good fortune to kill a reindeer and a mountain goat. The mountain goat is a very beautiful animal, brilliant white, with spiral horns, but shy and swift as lightning, so it is not very easily caught. When the feast was over we went to fix up our tent, and the hunters started off in another direction. We spent most of the rest of the day in feasting on the fresh meat we had pur chased. From the beginning of my career as a sailor I had noticed that the rations dealt out to us were much too small for a man to do any real hard work on, so I always utilised every opportunity, and I did so now, of laying in as good a foundation as possible to make up for the shortage in the days to come. Jimmy had made the same observation as regards his inner man, and had made the same plans as I had. Like all women folk, Kappa ate but little, and Mogg had heartburn and could not eat anything. We had a good deal of trouble with the dogs, who every now and then began to fight with the strange dogs, so that we had to go out and separate them. In the evening the hunters came home with two reindeer, and said they had seen a herd of sixteen. They told us that the deer generally remained in these regions throughout the whole of the winter. We bartered for a large joint, which we cut up into small pieces and took with us.

The next day we had a good stiff climb to get over the ridge of land; it was steep and full of roots and stumps, but fortunately they were not very large, and after a good deal of hard work we again reached the river ice. By 11 o'clock in the morning we were at Blow-hole, a notorious narrow pass between rocks 1,500 feet high—the very mention of which causes a shudder. The ice was all strewn with pieces of stone blown from the rocks, and it would have been no laughing matter to have been hit by one of these flying pieces. It was blowing so hard that I had to lie down flat on the ice, and the dogs rushed helter skelter with the sledges. Those acquainted with the region said it was very mild weather for the place, so I had to be grateful that it was not what they called really bad.

Some time after we met an Eskimo family with two toboggans. The man was a special friend of Jimmy and known as one of the cleverest hunters around Herschel Island. He had no less than sixty reindeer lying spread out in the fields and was now on his way to Herschel to get help to cart them in. It was only 1 o'clock in the afternoon when we met these Eskimo but we agreed to stop and pitch our camp together and partake of their reindeer. Jimmy and I winked at each other behind our leader's back, happy at the chance of having another good meal. This was the second time we made a voluntary stop in the middle of the day, but I made no objection, as the river was open in several places further up and so we were in no hurry. Both man and beast required to be properly fed and we could make room for a little more meat on the sledges. Mogg had taken a whole sack of tea with him and this proved very useful, as the Eskimo here would sell their immortal souls for a pound of tea.

As we advanced towards the south the landscape assumed a milder character, the rocks rounded off and sloped uniformly down towards the banks of the river. Here too the snow was firm and gave a good foot-hold for us as we ran in front of our sledges. We crossed a great number of reindeer tracks and now and then the track of a wolf. The latter prefer more southerly regions when there is food enough, and this was the case this year. When we occasionally encountered water on the ice we pulled on our water-proof boots and managed to get along fairly well. On the 30th we reached the source of the river, a large lake surrounded by high mountains; I estimated the highest of these at about 4,000 feet, and here the Eskimo family we had recently met had established a depôt. In these regions they establish their depôts on quite a different system to that adopted by our Nechilli friends. They erect a platform on four standards about the height of a man; they then lay the food on this and well cover it with pine branches, so that Mr. Fox can come along and sniff and jump as much as he likes; he gets about as much taste out of it as out of the famous sour grapes.

Mogg showed me a hill at the foot of which a tragedy was enacted some years ago. A number of the crew of the whaling fleet had conspired together and deserted with the sledges laden with stores, weapons, and ammunition. Some officers were sent out with a number of Eskimo to overtake and arrest the deserters, and it was just at this spot they were discovered, as they were building snow huts for themselves. They were commanded to surrender, but answered by opening fire, and a fight began. Two deserters were shot, two others surrendered, and the rest fled to the woods. One would have thought that nothing but certain death awaited these fugitives, here in the midst of winter and without either food or clothing, yet five of them, after terrible sufferings, reached Fort Yukon; the rest, however, perished on the way.

We crossed the water and pitched our tent on the other side. The following day a walk of two or three hours brought us to a pine-covered headland, where we halted and changed our sledges for the toboggans, and commenced transferring the loads. I thought I should never find room for all my load on that little toboggan, nor should I have done so, if I had had to do the loading myself, but Jimmy managed it. There is an art in packing a toboggan. The load must not be too high, or it will turn over; it must not be too broad, or it will project at the sides and act as a brake; consequently the load must be packed low and narrow—not a very easy job when one has much to pack, and, besides, the toboggan should be slightly back-loaded. We stood the sledges up against a couple of trees, and also left behind us a quantity of other articles we found we could dispense with, intending to collect them on our return journey. When on the river we had crossed the borders of Canada, and were now in Alaska. We completed our work early in the afternoon, and were able to enjoy a little rest in the tent. It was a very pleasant evening; we spread the floor of the tent with fresh pine branches, and the dried wood burned brightly and cheerfully in the stove. Some pot or other was always over the fire; we could never have too much water. We had a pack of cards with us, and Jimmy and Kappa were enthusiastic players. They knew a great many different kinds of games of which I understood nothing, but they grew so enthusiastic over them that they screamed and roared like children. When it began to get warm, and it varied now between 86° and 100° Fahr., it was always both desirable and pleasant to take off one's skin clothing to prevent it getting wet. For decency's sake we kept our shirts on, though we should have preferred to dispense with them, but we had to consider Kappa, as being one of the fair sex. The lantern was suspended from the middle of the roof and gave a very cheerful light; Mogg and I wrote up our diaries. One thing Jimmy and Kappa taught me, which I had never practised on a sledge trip before, and that was to wash myself every morning, and, if I forgot it, Kappa would at once appear with soap and water. Strange to say, they could not imagine anyone beginning the day without washing.

At 8 o'clock the next morning we continued on our way, but Jimmy's knowledge of the road ended here; however, we pushed on very confidently. The mountains in front of us were given on the map as 9,000 feet high, but with all due respect I take the liberty of estimating them at not more than 5,000 feet. On November 3rd we stood on the summit, which forms the watershed between the rivers flowing south and those flowing towards the Polar seas. The country was all mountainous here, and the wooded districts lay a little lower down on either side. There could not have been much wind on the mountain, as the snow was loose and deep, consequently the work of the dogs was very heavy. My dogs especially fared badly, as my toboggan was very roughly made of spruce wood, which soon splintered up underneath and turned into a sort of harrow; but the other one, made of birch, was as smooth as ice. Our marching order was as follows: Kappa and Jimmy took the lead, to show the way and make a track for the dogs, then came Mogg, who acted as a steam-roller, as he floundered about and made a splendid track. They travelled on snowshoes and made a track just wide enough for the toboggans; then came the dogs with the first toboggan and I followed with mine. I now saw the utility of the tackle they generally use, whereby the dogs are harnessed up in single file and are forced, whether they like it or not, to keep to the track, and this is of very great importance to those who follow.

When up on the hill we looked down into a little valley and saw that it led to Porcupine River, and once there we would be all right. There was a steep descent to the valley, but the snow was soft and I anticipated having a good, pleasant slide. I unharnessed the dogs, stretched myself out on the toboggan, and let her go. But I had reckoned without the dogs. When they saw the toboggan start they ran forward to get into their places again; they got in front, but they did not find their places. Down the mountain side we went altogether —toboggan, dogs, and I, one over the other, till we stopped at the bottom of the valley. For the most of the way I had had the dogs and the toboggan on top of me, and was mad at the stupid creatures for having spoilt my slide. I got up and brushed off the snow from myself and saw Mogg, who had found his own way down, standing a little distance off, splitting his sides with laughter. On the summit of the hill, Kappa and Jimmy were screaming with delight. I thought of venting my wrath on the dogs and giving them a good thrashing, but eventually decided to join in the general laughter. The Eskimo, with their heavy toboggan, were wiser than I. They took hold of it on each side and let it down quietly.

The little valley led first towards the south-east, then due south, and eventually it led again into the mountains; so it did not extend very far. The sun shone brightly in the mid-day sky right in front of us, showing us the way. By going straight towards it we should pass over the mountains to Porcupine River. In cases of this kind, the Eskimo are invaluable; they at once see, from the general lie of the land in which direction progress will be easiest. This time, however, there was a difference of opinion between Jimmy and Kappa, but Kappa's arguments seemed to be the most convincing and Jimmy at last yielded to them. Later, however, we found that Jimmy was right, though Kappa, of course, would not admit it.

It began to be very cold up here in the mountains. We had no thermometer with us, but judging from the drift snow I calculated the temperature to be considerably below -22° Fahr. We now started in the early morning when it was dark, the Aurora Borealis every now and then lighting us on our way. I often regretted now that I had not brought my snow shoes with me, as ski often cut into the deep snow and catch in the twigs or in the large grass tufts; even toboggans are not very practical on this kind of ground as they are constantly turning over and causing much annoyance and discomfort. Finally on the 4th we reached a real river bed. Certainly it was not very wide, merely a brook, but sharply defined with high banks. And here we travelled splendidly over the ice. The dogs went at full gallop. However, the brook soon twisted and curved so much that we had to take short cuts by crossing the land, and on the following day we came upon a broad river bed. We found out later that this was Coleen River, one of the many tributaries of the Porcupine River. Here it was excellent travelling, there being a couple of inches of snow on the ice, and I thought I would show them what could be done on ski. Mogg, who had to trudge on snow shoes over the mountains, because the dogs were not capable of pulling him along, now took a seat on my toboggan; and on a track like this, my dogs could have dragged double his weight. So off I went. Jimmy was leading, but the snow shoes did not glide along like the ski, and I quickly passed him. "Well Jimmy, what do you think of ski now?" I was soon a long way in front. Travelling over the ice was not altogether without danger as the river was open in places; these could always be seen from a distance, but sometimes the ice is so thin that it is very easy to drop through if you are not very careful. Now the high peaked mountains gradually disappeared behind us and we came into large woods.

The first two or three hours in the morning were always the hardest both for man and beast—especially for the latter—as they were stiff from the previous day's work and rather lazy, but Jimmy soon whipped them up and made them lively, and then we went along smartly. We had, however, been compelled to reduce the dogs' rations and the effect of this was soon apparent; they grew thin and lost strength. We ourselves had still provisions for some days, assuming that we used them as carefully as before.

On November 7th, at 3.30 in the afternoon, Jimmy suddenly stopped; his sharp eyes had discovered something unusual away on the ice. He rushed to the spot and called out to us, "Itkillich tomai!"—tracks of Indians! His voice had an echo of gladness in it. Now our troubles would soon be over and we should have plenty to eat. We followed the track and soon came to a wooden hut. My excitement was intense, for I was at last to see real Indians, who, in boyhood's days, had so filled my imagination with vivid scenes. I expected to see the door open and a copper-coloured fellow emerge, with feathers in his hair, swinging his tomahawk over his head, and yelling "Ugh!" to us. Or perhaps he was lurking behind one of the trees in the wood. The door opened, and out came a quiet man in black clothes and wearing a black hat. He stood quietly and looked at us. We greeted him in a friendly way in English, and he also answered us amiably in the same language. Shortly afterwards his wife appeared. It might all have been an incident from a walking tour at home; they looked exactly like a couple of peasants from the Norwegian highlands. We remained with them for a couple of days, and fed up the dogs as well as ourselves. They sold us some frozen fish and some elk meat in exchange for tea and candles.

On the 10th we pushed further along the same river course. Birch trees began to appear here and there, and many other things showed that we were getting further south. On the afternoon of the 12th we saw the track of a toboggan and of snow-shoes. We followed these until dark. We pitched our camp and followed the tracks the next morning, but lost them owing to fog setting in. At 10 o'clock in the morning, Jimmy told us he could see a wooden hut on the bank; his eagle eye had not deceived him, and when we reached the hut we found two women there, but the fancies of my boyhood received a rude shock; surely the squaws of the brave Mohicans, or even those of the sly Iroquois, could never have looked like these! One of the women had the under lip hanging right down to her breast, and the other held her head on one side and regarded us with a very surly look. They were two horrid guys. The meeting between them and Kappa was very effusive. They greeted one another and chattered together as only old cronies can, and none of them understood a word of what the other said, not a jot. We got a bundle of dried fish from them in exchange for some tea and biscuits. The husband of one of the women and the son of the other had gone away two days before to see a merchant at Porcupine River. The friendly ladies explained to our great satisfaction that by following the tracks of the men we could take a short cut and save two days' journey. The one with the surly look accompanied us to put us on the track. It was difficult enough to clamber up the hill to the wood where the track was ; repeatedly we had to carry the toboggan bodily, but once on the top the track was splendid. Our guide left us and we started off at a good speed. Here, again, I found further cause to praise the harness more than ever. If the dogs had had separate traces they would certainly have run one on each side of a tree and brought us to a standstill ; but now running the toboggan in a smoothed-out track was mere child's play, and there was no need for me to be in front of the dogs to encourage and lead them. Everything went automatically. So I pushed on to catch up with the Eskimo. Mogg lay face downwards on his load singing and humming. Everyone was in good spirits at the idea that we would be in Fort Yukon in a week. When I overtook the Eskimo I heard the quiet Jimmy shouting for joy whilst Kappa held on to the back of the toboggan and danced with delight.

Chapter XII.

Towards the evening of the 14th the dogs got scent of man and food, and they began to run as they had never run before. It was rather steep going down, but it was impossible to hold the dogs in. I was in front on ski, and every curve and ledge I whizzed over made me shudder for the loads behind me. Finally I reached Porcupine River near the little Indian colony, where lived the merchant already referred to. Poor Mogg arrived last. He had had a most anxious time holding on to the load, and in reply to my question declared he had had no time to admire the beauties of the wooded landscape. In order to prevent any fighting between our dogs and those of the Indians we set up the tent at some distance from the colony. The merchant turned out to be an Indian. He was a very fine fellow, about six feet high, with dark hair and full moustache. He was dressed in black and wore a white fox-skin collar. His stock of goods was not very varied,—some dry salmon, that was all. He thought we could cover the rest of the way to Fort Yukon in four days. We bought some salmon from him, and both we and the dogs had a good meal. The next day we left the brave John Alvert, such was his name, and continued our journey.

Had we known it we could have considerably shortened our way by now and then cutting across the land, but not knowing this we had to follow the twisting bed of the river the whole time. We passed several log huts which were uninhabited. There must be a great number of hares in these regions as the snow was often trampled hard with tracks, and occasionally we found a dead hare; probably we had disturbed some bird of prey at meal-time. When the dogs got scent of such a tit-bit they started off at full speed; of course it was only the first who got the prey, but this never taught the others anything. They rushed ahead each time with unabated energy and renewed hope. Even stout Fix, who was the last dog in the team, ran as if he would break his harness. Fix was one of the dogs Atalanga had brought to Ogchoktu, when he brought our first mail. But later on during our stay in Eagle City, Fix got so fat that I had to leave him behind; he was incapable even of following the sledge.

On the 18th we again saw fresh tracks, and as they led inland we followed them. In the evening the dogs began to get very lively, they must have got scent of something, and that something quite out of the common, judging from the rate at which they dashed along. At 5 o'clock we sighted a house, and half-an-hour later we were with the Indians in Salmon Creek. We occasioned quite a stir when we arrived. The first thing we noticed was that there was only one man and a number of women. The fact was that all the men had gone to Fort Yukon to trade; "Old Thomas" alone remained behind. He invited us in and told us he had several times travelled the same way as we had and it eventually transpired that he and Mogg were old acquaintances from Herschel Island. Exceptional hospitality was shown us;

the one room which was already occupied by five people was vacated and placed at our disposal. There were two stoves in it, and these were kept continually going. Old Thomas was a remarkable fellow. He spoke four languages, English, French, Eskimo, and Indian, and had a great deal to tell us about his many wanderings. Mogg, who was more kindly disposed to Eskimo and the Indians than to his own people, presented him with a little of everything, tobacco, tea, matches, etc., and when we started the next morning, the old man declared that Mr. Mogg was an angel. I did not say how far I agreed with this opinion.

While we were made perfectly welcome, the dogs had made themselves at home very unceremoniously. Goodness only knows how, but somehow they had attacked the depôt and helped themselves to its contents so freely that in the morning they could hardly move. We should really have reached Fort Yukon that evening, but owing to the overfed condition of the dogs, they could not get along very quickly and we had to pitch our tent for the last time. The next morning we met four Indian tobogganers. These were men of the colony returning home. They looked very smart in their bead embroidered clothing. The harness on the dogs was also embroidered and adorned with little bells. Indeed they regarded their clothes as of great value, and once when I inquired the price of a jacket, I was asked thirty-five dollars for it; but clothes are generally dear in Alaska.

At 1.30 in the afternoon of November 20th we arrived at Fort Yukon, which is situated on a steep river bank, where the Porcupine joins the Yukon River. I cannot say that the fortress made any great impression on me. Two white merchants lived there, and I must mention the excellent and very kind Mr. Jack Carr; otherwise the colony consisted of some thirty odd Indian huts. The business of the merchant is to barter for furs with the Indians. There is also a school and a mission here. I had reluctantly to take off my fine Nechilli dress, as it was the object of too much attention on the part of the numerous youths, who followed me in large groups wherever I went.

Glad as I was to arrive, I was disappointed with Fort Yukon. I had built my hopes on finding a telegraph station here, but, unfortunately, the nearest station was Eagle City, 200 miles to the south, higher up the river. There was no help for it. I wanted to get into communication with home, and to do that I must go on to Eagle City. Jimmy and Kappa remained behind. Kappa was fairly worn out after the journey and wanted rest. Mogg and I therefore engaged an Indian as guide, this being still necessary at that time, in the narrowest portion of the Yukon River. The river is full of islands, forming a network of sounds and channels which one must be acquainted with to find one's way through, and the mails were not regular enough to make a sufficiently reliable track. The mail leaving Fort Yukon has its terminal point in Circle City; there it is taken over by another mail carrier, who takes it further south. But the whole of the postal communication between Fort Yukon and Dawson City, *via* Eagle, is effected by four carriers, who use sledges and dogs. Between Eagle and Dawson Cities horses are also used. The total distance from Yukon to Dawson is estimated by the carriers at 300 miles.

I arrived at Yukon with only about half of my toboggan. Two of the four boards were worn out, so I had to buy a new one from an Indian. Our load was light now, as we no longer needed tent or other equipments. The little we carried with us was on my toboggan; the guide, Charlie, had his own toboggan, and Mogg travelled on his. We started at a fine speed up the river, and it was evidently Charlie's ambition to show us what sort of people the Indians were. He was ahead, and ran in front of his dogs as hard as he could. However, what with the light load and their long training, my dogs were not to be beaten; they kept well up to Charlie's toboggan, and I followed behind on my ski. Mr. Charlie out-distanced none of us. In the evening we came to a wooden hut, put up for the mail carriers. It was very comfortable and cosy, and divided into two rooms, one for the toboggans, and in the other were two bedsteads, two chairs, a table, and a stove. The bedsteads were covered with fresh pine branches, and were very attractive after our twenty-miles travelling. The next day we passed a little hut, where a woodcutter lived. Of course, we looked in. Mr. Lee Provost was quite an exceptional fellow. He seemed to be endowed with all the good qualities of mankind, and his personality and open hospitality made such an impression on us that we remained the night with him.

On the 26th we arrived at Circle City and said good-bye to our guide; and we were not sorry to do so, as he was an impertinent and conceited fellow, whose principal desire was to show us that he had learned from the

LEE PROVOST'S HUT.

missionaries that white folk and coloured folk were equals. Owing to this he managed to make himself very disagreeable, showing that good teaching may sometimes turn out badly.

Circle City must be regarded as "quite a little town," as witness its liquor shops and dancing saloon, to say nothing of the fighting and drunkenness resulting therefrom. We were lucky enough to find out that the mail-carrier, Mr. Harpar, was starting southward next morning, and, of course, it was a great advantage to us to accompany him. The mail-carriers in Alaska are splendid sledge drivers; their dogs are of the best, but very different to the Polar dogs. As a rule, they are

MOUND OF EARTH CONTAINING 1,000,000 DOLS. WORTH OF GOLD. NOME.

short haired and long legged; their long legs are an advantage in the deep snow, and, as they sleep indoors, they do not need any thick coat.

From Circle City, towards the south, one meets with the so-called Road Houses, small log huts providing "food and lodgings for travellers." They are situated along the river at intervals of about twenty miles, and generally consist of three rooms, the room for the guest, the kitchen, and a little room for the proprietor. All arrivals are packed into the first room. Those who have not their own beds with them must share with another; but people in these parts, after travelling all day and arriving very tired, are not very particular. For us, who had come from the northern regions, these "hotels" were perfect wonders of comfort and elegance, but they were also very expensive. The sleeping place, whether you slept alone or shared it, cost a dollar. Each meal, 1·50 dollars. One day's board and lodging, with three meals, came to more than 20 Norwegian kroners (22*s.* 6*d.*). But the prices of everything are exceptionally high in Alaska, and, when gold is discovered in the neighbourhood, they go up by leaps and bounds. These high prices are naturally occasioned by the difficulties and cost of transport. In Fairbanks, on the Tanana River, where the last great discovery of gold was made, a pair of snow-shoes fetched forty dollars, and a dog was worth fifty dollars. Even these prices are insignificant as compared with those which were paid on the occasion of the rush to Klondyke. I have been informed, on reliable authority, that 2,500 dollars were once offered for a team of five dogs—and the offer was refused. Everything else was on the same lines.

I now approached Eagle City with great excitement. At last I was to be in direct communication with home, and get all the news from my own fatherland. As we rounded the last point, there, only two miles away, lay

Eagle City, with its blue smoke standing out darkly against the bright sky. You can imagine how overpowering is the thought that within a few hours you will be in touch with the dear ones at home. When we got near the town, we left the ice and drove into the city to the telegraph office, which was situated within the walls of Fort Egbert. Fort Egbert was occupied by two companies of the 3rd Infantry Regiment, and the officers received me most kindly. I despatched my precious telegram, which was only just in time, as shortly afterwards the wires broke down, owing to the intense cold.

I remained two months in Eagle City, waiting the mail from home. I shall never forget that time, as it is associated with some of my most cherished and pleasant recollections. I was the guest of Mr. Frank N. Smith, Manager of the large firm of the "Northern Commercial Company." We have a saying that "when three days old, guests and fish get stale," but this saying could never apply in Mr. Smith's house, and now, after the lapse of time, while writing these lines, my mind is still filled with the warmest and most heartfelt gratitude to this household.

On February 3rd I again started for the north. I had now received all the news from home, including letters and papers, and was glad to get away.

On my way back I again met with liberal hospitality in Alaska. Mr. Jack Carr, of Fort Yukon, whose guest I was for three days, did all in his power to facilitate my journey over the mountains. I also owe a debt of

FORT EGBERT, NEAR EAGLE CITY, ALASKA.

gratitude to Mr. Daniel Cadzow, of Rampart House, on the Porcupine River, the last point of contact with civilisation, where I passed some pleasant days before I started back for the mountains. Nor must I forget to thank the two brave mail-carriers on the Yukon River; they always did all in their power to help me, and were exceptionally kind.

I then started for the north. The whip cracked, the dogs pulled, and we were off northwards—to the "Gjöa" and my comrades.

CHAPTER XIII.

CONCLUSION.

THE first vessel to enter the open channel off the shore, was the "Bowhead," Captain Cook. She anchored off Cape Sabine to pick up driftwood, there being a large quantity at that point. When the other two vessels saw that the "Bowhead" had succeeded in getting round into the open channel, they followed, but were unable to get in until the afternoon. Looking across the ice, we could see they were the "Alexander" and "Jeannette," of San Francisco. As I had to see Captain Tilton before my departure, with a view to arranging matters connected with Dr. Wight's transfer from the "Alexander" to the "Gjöa," we lay there for the time being, and waited. We had enough to do as it was. During the last few days Lindström had baked a large quantity of white bread and stowed it away in casks. By this means we should have soft bread for some time to come. His oven in the house on shore was still full of bread, and we could not leave until our "Handy Man" returned. We utilised the remaining hours collecting wood, in addition to what we had already. We stowed wood in every possible and impossible nook and corner of the boat. We were lying on the starboard side of the wreck, quite lightly moored and ready to get under way at any moment. It was a beautiful afternoon and the sea was perfectly calm. The dogs were lying on the forecastle, basking in the sun. Had they anticipated what was impending, they would scarcely have been lying so quietly. We had become fond of our swift animals, notwithstanding their vices and all the trouble and anxiety they caused us.

The Eskimo had assembled on board the "Bonanza" to see us off; they were all there, Kataksina, with little Anni on her back, as well as Kunak and Neiu with their wives and sons. We were now waiting with growing impatience, there was nothing more for us to do. At last, the two stout friends came upon the forecastle, each carrying a loaf of bread. Sten intended going out with us to meet the "Alexander," and returning with her. It was an anxious moment when the two corpulent gentlemen stood poised on the narrow gangway; we looked on with our hearts in our mouths, trembling for the safety both of themselves and their loaves; then the command was given: "Cast off! Full speed ahead!" And the good ship "Gjöa" started on the last and final stage of her long voyage. As we passed Wiik's grave, we lowered our flag to half mast and paid a last tribute of friendship to his memory. Then the flag was run up again, and we ploughed ahead.

The beach presented an animated scene. All our good friends and acquaintances, both white men and Eskimo, were there, busy collecting drift wood, and they kept cheering and waving so that we had to dip our flag again and again in response. Meanwhile the "Alexander" was drawing near and I gave the order to "Stop," so as to wait for her. As the vessel approached, Captain Tilton

WE LOWERED OUR FLAG TO HALF-MAST—THE LAST TRIBUTE FROM HIS COMRADES.

shouted to us several times, but we could not understand a word. He must have been in a hurry, as he did not stop, but continued his course towards King Point. Having, as I have stated, promised Dr. Wight to take him on board, we could not do otherwise than turn round and follow her. It was no great distance after all. The "Alexander" and the "Jeannette" were lying to, and when we got alongside we also lay to. As soon as we had put Sten ashore and had taken the Doctor on board, we hastened to resume our voyage. Off Cape Sabine we stopped and sent a boat over to the "Bowhead," still lying there, to get some sperm oil. It was Dr. Wight who had advised this, on account of Manni who had been unable to get rid of his cold.

After we had been under way for a couple of hours or so, the engineer came and reported that the whole engine-room was flooded with water. Heigho! it must be the effect of the water on the vessel, she must have sprung a leak. We at once sounded the pumps but found that she had no more water in her than usual. Eventually we found that the sole cause of the mischief was some old ice we had on board; it had suddenly melted and formed a channel communicating between the main hold and the engine room, all the water rushing aft. We escaped this time with nothing more than a fright. The engine worked well and we made three knots. A slight puff of wind from the east with a swell, indicated that the bay at the mouth of the Makenzie River was open. At last we got round to the outside of the ice and set our course towards the west. The wind freshened, and the swell was heavy. When I came on deck at 6 A.M. next morning, we had King Point on our beam abaft. I took up my telescope and singled out the familiar spots, the wreck, the houses and the cross. This cross rose high up against the clear sky, and seemed to give us a message of "Love to all at home."

It was blowing a fresh breeze from the south-east, and we went smartly ahead westward. When we came to Herschel Island we found it surrounded with ice. But what did that matter to us? We had no business on the

SUMMER AT KING POINT.

island, and laughed at the sapient skippers whose last advice to us was that we should run into the harbour and remain lying there quietly for at least a fortnight; but now we could see ahead of us, beyond the island, that the condition of the ice was most favourable, and we laughed at the sages. But, alas, it is not well to disregard the voice of experience. We had not proceeded far on our beautiful course when we found that the open water was merely a kind of inlet in the ice to the west, a mere *cul-de-sac*. We had to return promptly and try to make our way into the harbour. A fresh westerly wind opportunely laid the harbour open for us, but we had to run so close in under the north-east point of the island that we had barely nine feet of water. However, we managed to squeeze through, and at 2.30 A.M. of July 13th we cast anchor.

This afforded us a very good, though brief, opportunity of studying Herschel Island without its pall of snow. The island is small and is covered all over with moss. It is separated from the mainland by a narrow, shallow sound. When Franklin passed here in 1826, he did not observe the excellent harbour on the east side of the island. Hence he says that the narrow sound is the only place of refuge for a vessel between this and the Mackenzie River. Franklin did not at that time anticipate the great importance Herschel Island was destined to assume. The harbour is good in all winds. It is true that a south-west wind would blow right into the harbour, but it could hardly do any damage, although it has happened, in very exceptional weather, such as is not experienced more than once in a generation, that boats have been driven ashore. It was Franklin who discovered and named the island, but it is the bold American whalers who have made it what it is. They were exceedingly badly off for harbours on the North

American coast, and the discovery of Herschel Island was therefore of great importance to whalers. It was in 1889 that they pushed through to this island. It had taken many years to get so far, and a good many lives were lost in the attempts. The first calamity that befell the struggling whalers' fleet was due neither to foul weather nor to ice. During the war between the Northern and Southern States in 1865, a man-of-war belonging to the Southern States went north and burnt thirty ships and destroyed about three million dollars' worth of property. In 1876, thirty vessels got stuck fast in the ice off Barrow Point. Some seventy men of the crews abandoned their vessels to save their lives, but nothing was ever heard of any one of them. Later on, in 1897, a number of vessels were crushed by the ice. Lastly, in 1905, the ice set in towards the coast a month earlier than usual, stopping all traffic. But no vessels were lost on that occasion, as all managed to find a harbour of refuge. On the whole, the hunt for the precious bowhead whale has been a very costly one.

Of the bowhead whale, the whalebone alone is used; all the rest goes to the fishes. But then the present average value of the whalebone taken from one whale is 10,000 dollars. Whale hunting is not by any means easy or free from danger. The bowhead whale is very wary, and is scared away by the least noise. As soon as a whale is sighted, the propeller is stopped and sails alone are used. While the whale is still a long way off, a boat is lowered to begin the actual hunt. Oars must not be used; it must be propelled by sails only. The little boat is steered direct to the huge monster, the harpooner standing in the bows ready to throw his harpoon. Shooting is out of the question, as a shot would scare away all the whales for miles around. Tonite is used as an explosive. If the whale is not killed by the first throw, he darts off madly, and ample rope must be given him if the harpooners are to follow him, just as in hunting the bottle-nosed whale between Jan Mayen and the Faröes. If there is ice in the fairway, the hunters must be on the alert. If it becomes necessary to cut the harpoon line, this means an absolute loss of something like £1,800. When killed, the whale is towed to the vessel. The head is cut off and taken on board, and the carcass is sent adrift. Then the whalebone is taken out and the head is hove overboard.

The first bowhead whale, *Balena mysticetus*, was caught in the Behring Sea in 1843. Five years later the first whaler passed through Behring Strait, and was soon followed by many others. In 1905 the whaling fleet comprised fourteen vessels, all, except the "Bonanza," being fitted with auxiliary engine power. This year, the brig "Jeannette" scored the largest catch, eleven whales. For sixty years this trade has been carried on at enormous profits, but attended by grave perils and by the loss of many a life. All these lives and all this property are risked year by year to supply the markets of the world with the whalebone of the bowhead whale. I inquired what this costly material is used for, and I learned that it is chiefly used for the manufacture of corsets. A ladylike figure is an expensive thing; but I think that, after my experience as a Polar resident, I would vote in favour of dress reform.

In spite of the early morning hour, all the Eskimo in the harbour were astir. After the departure of the whaling fleet, we were the only "lions" in the place, and

VEGETATION AT KING POINT.

were treated with the greatest deference. After some hours' sleep, we went up to the highest point of the island to survey the ice conditions. There was much ice lying to the westward. Along the coast there was an open channel in the ice, but it was impossible for us to judge from where we stood whether it was wide enough for us. At any rate, it would be a great advantage to get into the open channel, so as to be in a position to make the most of any opportunity that might present itself. The only means of access to the open channel was the narrow sound between the island and the coast. We had heard many contradictory reports as to the depth of the sound, but, to make certain, Lieutenant Hansen went out into the sound, accompanied by Hansen, the Doctor, and Foss, with an Eskimo to act as pilot. This pilot did not prove to be of much use. Each time he took soundings and found the water shallow he simply called out, "Water very small!" which was intelligible enough, but afforded little information to those abaft. The results of the examination showed that the bottom was very unequal. Even if it were possible to thread one's way by twists and turns along a navigable channel, it would, after all, be too risky, and we decided to bide our time. On the way back, the party met a number of Eskimo who had had good hauls in fishing, and we bought a large quantity of fresh fish from them.

We now arranged that one of us should go every day up to the highest point of the island to inspect the ice conditions. The highest point lay quite over to the west of the island, a stiff walk of close on two and a half miles, over very difficult ground. But, as for the vegetation on this island, King Point was a desert compared to it. It was, so to speak, carpeted with flowers, and Lindström was in clover. Early and late he might be seen with his green botanical collecting-box on his back, and he always brought some rare specimens home. The richest flora was found at the back of the cemetery. The burial-place on Herschel Island was divided into two parts: one for whale hunters and one for Eskimo. The whale hunters' graves were all well kept and adorned with painted crosses. But the Eskimo's presented a most remarkable appearance. It was as if a tradesman kept his store of goods there, for the Eskimo put their dead into ordinary wooden boxes and deposit them in

ESKIMO GRAVES AT HERSCHEL ISLAND.

rows on the bare ground. Only very few had placed these boxes on wooden trestles; the majority lay on the ground. I could not help wondering how they were able to identify the remains of those dear to them.

When the whale hunters arrived at the island it was inhabited by about 500 Kagmallik Eskimo; now there are only very few, and the great majority of these are a mixed race. They live in small wooden huts; very unhealthy as far as I could judge. Besides, there are a number of large storage sheds and small depôts. The former missionary's house is now occupied by Major Howard and his staff. This man has the by no means light task of maintaining order among some hundreds with only one man under him. He also has to collect

WHALERS' GRAVES AT HERSCHEL ISLAND.

duty from the Americans who, of course, are on Canadian territory here.

Manni went out assiduously to shoot ducks; these are to be found in large numbers. I forbade him to go inland as I did not care to have him infected with any of the Eskimo's various diseases with which civilisation had gifted them. For instance, syphilis was very prevalent among them. But what strikes the stranger most is the rising generation here. They bear the stamp of a very varied admixture of races; a pure Eskimo type is exceedingly rare among them. The children differ greatly, not only in facial type, but also in their dress; for instance, I saw a little girl with a red frock, black shoes, and a "baby hat," which suited her

ESKIMO HUTS AT HERSCHEL ISLAND.

admirably. Surely she was not an Eskimo? Then the mother came upon the scene, and, though not a full-bred Eskimo, she was certainly a half-breed. The child was the third generation of race admixture. The mixture of Eskimo and negro has a most ludicrous effect. I cannot find a name for the specimen I saw, but it was exceedingly comical.

The Eskimo on Herschel Island have grown so accus-

TUPSI. ESKIMO WOMAN AT HERSCHEL ISLAND.

tomed to white men's food that if they cannot obtain it they deem it a great hardship. If the flour supply runs short they feel the want of it very severely, and this year even the white men have been so badly off in this respect that there has been little to spare for the Eskimo. Just at the time I refer to the poor people were waiting the arrival of the tender of the whaling fleet, which was to bring food for all. The tender did not get further than Point Barrow, where the whalers met her and collected the supplies ordered; so the Eskimo were bitterly disappointed this autumn.

Although the missionary was away, religious services were held every Sunday. An old chief named Tomachsina conducted the service and Dr. Wight played the organ.

Here we came across the first vegetables of this region, "Kagmallik Potatoes" as the whale hunters called them. These are the roots of *Polygonum bistorta*, and do not taste badly, either raw or cooked. In the Eskimo language they are called "masku"; they resemble carrots with a skin like a potato, and have a slightly sweetish taste. The Eskimo gathered sackfuls of them and sold them to us.

On July 20th the wind was north-easterly, and all the whale hunters had told us this was the most favourable wind for driving the ice away from the land. It soon increased to a gale. But the daily look-out returned and reported that the ice was closer in than ever. On the 21st I rowed ashore with Ristvedt, Lund, and the Doctor to see whether this wind had in fact produced any effect on the ice. We proposed to row a couple of miles along the east coast so as to find an easier ascent to the top. While on our way we met Manni, who had started out after breakfast, and we called out to him something or other, I do not remember what. We pulled the boat up on to the landing place and went up to our look-out. The ice looked promising; the north-east wind had begun to take effect. As we were approaching the boat on our way back, Lund suddenly exclaimed: "I think they have hoisted the flag on board. What can it mean? But they have not hoisted it right up to the top!" The telescope was brought out and we saw that the flag was half-mast high on the "Gjöa." There can be no more unpleasant sight than a flag at half-mast. We tried to console ourselves with the idea that possibly it referred to one of our Eskimo friends on shore. Yet we could not feel assured on this point. I at once thought of Manni, and I believe the others did also. We flew down the slope, left our boat there and made our way overland till we were just off our ship. We had already been observed and were taken aboard by a boat sent for us. We were right, it was Manni. He was drowned.

When we got aboard, the Lieutenant told me that while he was standing on deck talking to one of the lads, he saw Manni standing upright in his little boat, taking aim at a bevy of ducks. They were so much accustomed to seeing him under similar conditions that they took no particular notice of him. But a moment after when they looked that way, the boat was empty and water was spouting up from the sea by its side. Manni had fallen overboard. Quick as lightning Hansen and Foss were into another of the ship's boats, while the Lieutenant flew

MANNI (SUMMER, 1906).

up to the crow's nest to direct their movements. Barely five minutes could have elapsed when the boat reached the spot. But Manni had disappeared. The canvas boat was lying upright on her keel, full of water. The oars were drifting on the sea but the lad and his gun were gone. A heavy wave had struck the boat as he stood upright in her looking after the birds, and he had fallen overboard. He never rose again. Another sad illustration of the unfortunate fact that none of the Eskimo ever learnt swimming. I at once reported the misfortune to the Chief of the Police and asked him to provide for the funeral, if the body should drift ashore. But the Eskimo thought Manni's body would never be found as the current would carry it out to sea. It was a heavy blow to all of us to lose Manni under these sad circumstances, We had all become fond of him and were very anxious to take him with us to civilised regions and see what we could make of him there.

Next day I was again on our look-out. It seemed as if the ice was steadily continuing to recede to the west. But, after all, it might only be the same bay from which we had been obliged to return. As long as the north-easter was blowing right into it, it might be risky for us to try to get ahead that way again, and after consultation with my comrades, I decided to wait till the wind lulled. On July 23rd, at 1 A.M., the wind dropped, and we at once got all clear for starting. The anchor was weighed. There was still a little wind in the north-east, but it was quite sluggish. We followed the ice, which extended in a continuous mass from the south-west point of Herschel Island towards the west along the mainland. Landward of this ice there was an open channel along the coast, and what we had to do was to try to find a way into this channel. There were several inlets extending far into the ice. One of these extended to a distance of about fifteen miles from the harbour; there was only a narrow belt of ice separating it from the open channel off the coast, and it could not possibly be long before this barrier would be broken; but for the present we could not get through that way, and we therefore continued our westward course. The edge of the ice soon began to bear off towards the north, but that was not the course we wished to take. We went on until we all realised that we had again been trying to get through a bay in the ice, another *cul-de-sac*. So at 11 P.M. on the 24th we had once more to turn back to Herschel Island. We had to beat our way up to the harbour against the north-east wind, and it took us a long time to get in. We old tars took this calmly enough, but the new hands were sorely disappointed. The Doctor, particularly, was in a state of nervous excitement. At last, at 2.30 A.M. on the 26th, we were again lying at our old anchorage. Immediately after breakfast the "look-out" went ashore. He returned later in the day and reported great changes for the worse. We were now expecting a regular storm to blow from the land, but it did not come. After noon we sent Hansen on a tour along the east coast of the island to look out for Manni's body. But he returned without success.

Some of the bread Lindström had baked had got mouldy and had to be thrown overboard. The bread made with syrup kept well. Lindström was now the "Second Engineer," as also the Ship's Baker. Down in the engine-room we had fitted up quite a little bake-house, from which he supplied many French rolls and loaves. He had only one objection to establishing himself altogether in the engine-room: the "First Engineer," Ristvedt, "was so unreasonably fond of cakes."

On the 30th the first whaler entered the harbour. This was the bark "Belvedere," which had run short of food and fuel. On August 1st the look-out reported very favourable ice conditions, and I decided to weigh anchor at once and make another attempt to get out of the ice. We were under weigh at 4.30 P.M., and an hour later were overtaken by the "Belvedere," which passed us to the west. She was anxious to push on ahead, as scurvy had broken out among the crew. We followed the border of the ice. At 9 P.M. on August 2nd we moored to the ice seven miles from the coast and ten miles from Demarcation Point. We took advantage of the time we were lying moored to procure a supply of fresh water, of which there was an abundance in a large deep pond that had formed on the ice. We lay there all through August 3rd, quite hemmed in by ice. It was our King's birthday, and we hoisted the flag in honour of the occasion. We celebrated the day with such festivities as we could afford to indulge in; little more than a few extra beans in our coffee and a few currants in the wheat cake. We had nothing much else to make a feast with. But it may well be that our King did not enjoy his food any better on that day than we did, as we lay in the ice flying our flag in his honour.

The next night was rather an anxious one. The ice began to press, but it did not come to anything serious. The rudder was forced up a little, but it was soon set right again. About 6.30 A.M. on August 4th, the ice-pressure slackened and we got free. There was a fog, and the wind was in the west; the worst quarter we could have had it from. There was nothing left for us but to return to Herschel Island, and at 2.30 A.M. we lay, for the third time, at our old anchorage. The "Belvedere" had already returned, and knew just as much as we did, that ice was lying to the north-west. The schooner "Herman" had also arrived, and later, at night, the "Kaluk" arrived. The vessels began to collect, awaiting the arrival of the tender that was to bring fuel and provisions. During the two following days the barques "Treasure" and "Bowhead" came in, and now there were seven boats lying in the harbour. The whalers had decided to lie there till August 10th. If the tender did not arrive by then, they intended going to Barrow Point to meet her.

On August 9th the second mail from Edmonton and Fort McPherson arrived. It came by boat, and brought us the latest news. Of course, great sensation was caused by the news of the earthquake and fire in San Francisco. A Mr. Steffensen, who came with the mail, told us that a Danish expedition, under Mr. Mikkelsen, was on its way here, and that he was to join it. The object of the expedition was to search for land in the north. We were rejoicing in our intentions to search for land in the south.

Chapter XIII.

On the same day a gale sprang up from the north. At 1 P.M. the long-expected and anxiously awaited "Olga" entered the harbour. She had wintered in Minto Inlet, in Prince Albert Land, where they had met some Eskimo. Probably these were Kilnermium Eskimo, from the Coppermine River, of the same tribe that Hansen and Ristvedt had met on their sledging expedition in 1905. The "Olga" had lost several of her crew, and it was particularly unfortunate that both engineers had died, so that the crew could not get their engine to work. They had to depend entirely on their sails. The crew were two whole days out in their computation of time. They had seen a large number of whales, but were too few in number to make any catches. At 7 P.M. the "Bowhead" set out to find some of these numerous whales.

The north-east wind was still blowing fresh next day, and, being rather tired of Herschel, I decided to run into our old bay in the ice, to see whether the long-continued north-east wind had not opened a way for us through the narrow strip of ice we had observed on the previous occasion. After taking on board a large supply of driftwood, which the Eskimo, Manichya, had collected for us, we weighed anchor and ran out. A strong current to the west carried us out quickly. We held to the south-west all day long, in the hazy atmosphere, without being able to get a full view across the ice. At night the fog set in, thick as a wall. We very soon got from twelve fathoms into seven and a half fathoms, and thought we were not far from land. We found some ice here, lying fast to the shallow bottom, and we moored to it to await clear weather. We estimated that we were about two miles from land. Next morning we had to shift our mooring to some other ice, as that to which we had been moored had got afloat and was drifting away. At 6 P.M. the fog lifted and revealed a gladdening sight. We had got into the open channel along the coast, and had a clear course to the west as far as we could see. To the north the ice was still lying heavy and close. We therefore started the engine and got under way. It looked now as if the evil spell was at last broken. The open channel was at first very narrow, but became wider on the following day. However, the fog was so thick that we could not see much. A slight north-westerly breeze compelled us to tack, the engine not being powerful enough to make headway right in the teeth of the wind. Otherwise, it was just under such conditions that the engine did excellent service. If we had had to depend on the sails alone while beating up this narrow channel, filled with ice, and with a sluggish breeze, it would have taken us a wretchedly long time. When we were in danger of being caught in the ice, and everything depended on our rushing ahead in front of it, the engine was of the greatest assistance. And as this sort of thing happened continually, we were indebted to the engine for saving us much loss of time.

Off Icy Reef we repeatedly neared the shore and observed a large bright white surface, which looked like a lagoon or lake, but it could not be that, as the ice upon it would have melted ere now. It must have been the glacier near Icy Reef, of which we had heard on Herschel Island from a man who came from Camden Bay to buy provisions. It was not large, but, as far as we saw, it was the only one on the whole north coast. Moreover, the fairway around Icy Reef is full of ice-blocks and fragments of fresh ice. However, we had no time to go ashore and investigate the phenomenon.

On the 14th, at 10 A.M., we passed Manning Point, which is visited by many Eskimo. We observed there, in fact, a number of huts and stacks of driftwood, but no living soul. During the afternoon watch the ice was constantly forcing us towards the south. I did not like this, because Collinson describes Camden Bay where we thought we were at present, as being shallow and foul. In fact, we got into two or three fathoms of water there, but soon found ourselves out in open water again and able to go further over towards the north, where the water deepened quickly. At night the north-easterly breeze increased to a gale, so that we had to reef sails. Owing to the fog, the stiff gale, and more particularly to our uncertainty as to our actual whereabouts, I decided to go up to some ground ice and await further events. Suddenly land came in sight right ahead of us; we thought it was the Flaxman Islands, but, of course, we could not be certain.

The work on board was sometimes rather varied, there being so few hands. This evening, for instance, Lindström had to make the meteorological observations, take soundings, bake bread, and attend the engine. All the others were equally hard-worked, each in his own way. It is certain that our voyage would never have been accomplished had the men not been tractable and willing. In difficult situations we shared trouble and hardships in brotherly unity, and all rejoiced with one heart when difficulties were surmounted.

The ice to which we had moored was, in fact, pack-ice. It was full of large old hummocks, showing that it had not always been so still and quiescent as now. It appeared to me to be less in bulk and presumably newer than the ice off Greenland.

All the following day we lay still, owing to the fog. When it parted for a moment the Lieutenant at once took our bearings and found that we were actually lying three nautical miles north of Flaxman Islands. Next morning the wind lulled, and the fog cleared sufficiently to enable us to see each other. We then cast off, and went cautiously ahead under small sail and engine-power. We proceeded four nautical miles, constantly taking soundings. The ice was very manageable, and we made good progress. At 4 P.M. we passed one of the many small sand-banks lying along the coast. At 5 P.M. the ice became impracticable toward the west, and I decided to seek a passage inside the nearest bank. According to what the whalers told us, we should find sufficient depth. But the difficulty was to get through to the inner side. The whalers had told us we could find a way in between some of the banks, but in the fog it was quite impossible to distinguish one sandbank from another, so we were obliged to feel our way cautiously. We made everything clear to drop anchor rapidly, should it become necessary, and set our course inward. The bottom began to shallow up, and when we found nine feet of water we dropped anchor, so that we might explore in a small boat. Our soundings showed that this inlet was blocked. From the masthead I saw another inlet, and this was passable with a minimum depth of three and a quarter fathoms. Inside the banks we found quite open water, as the sandbanks prevented the ice from getting in.

At 5 A.M. we had Cross Island on our beam, where a cross had been raised as a landmark. At 9 A.M. we found ourselves in deeper water, from five to seven fathoms, and therefore were evidently out in the open sea again. We had gained a good deal by thus passing inside the belt of outlying banks; the ice on the outside might have detained us a long time. The waters about here are notorious for large accumulations of ice. The fog lifted very rarely. At noon we were right off one of the Thetis Islands, but could not very well tell which. To the west of the outlying banks the sea was clear and free from ice. Off Harrison Bay we again encountered ice, and were compelled to bear to the south. In the course of the night we travelled across the bay, and found we were in the "Pacific Shoal," with two and three quarters fathoms of water, under the lee of Cape Halkett. The ice lay close all the way and compelled us to keep very near in shore. In passing Smith Bay we had to keep to the inside, as, in fact, we had been compelled to do in passing all the other bays. The whalers had not sailed Smith Bay. We found it clear, and with an even bottom. The least depth we found was off Cape Halkett, but further in the bay, and to the westward, we found four and five fathoms. At 6 P.M. we sighted Cape Simpson and had thus got across the bay. The ice seemed to lie right up to the shore at the Cape; the whole east side was also filled with ice, but on the shoreward side there was an open channel, wide enough to admit us. We tried to enter it, but had to turn back as the water suddenly became very shallow, and we then made fast to some ground ice. However, we could only find a very small mass of ice to moor to, and the position was rather dangerous. The wind kept in the east all night, but the next day there seemed to have been a great change in the ice. We now made another attempt to get into the open channel along the shore, but an abrupt shallowing up from two to one and a-half fathoms induced us to retire with all possible speed. The ice on the seaward side was closely packed and again we had to make fast to the ground ice, which this time was very small, and the "Gjöa" had very little to depend on. At night the easterly wind freshened to a stiff gale. The ice lying to windward of us, which had been protecting us from the sea, now drifted away, leaving our little piece of ground ice at the mercy of the heavy rolling sea. The sky was overcast, it was dark as night, and the sea poured in a stream across the surface of the ice. At the same time a large quantity of loose ice got adrift, some of it floating towards the ground ice and some of it towards the boat. We endeavoured to ward off the shocks of its impact as far as possible by means of boat-hooks. Luckily we had got both ice anchors out, and they were needed. There was a great probability of the ground ice, to which we were moored, breaking up, and we were ready to drop the anchors at once should this happen. But in the darkness of the night it was not easy to keep the chains from being fouled by the drifting masses of ice. At last the day broke, and rarely have we welcomed daylight with greater joy and relief than we did after this terrible night. But before we could quit our unpleasant surroundings we had to get the ice anchors loose, and it needed a brave and active man to manage this on the small ice-floe to which the anchors were fixed. I selected Helmer Hansen for the work. He knew no fear, and was as nimble as a squirrel; as a rule he did at the first attempt what others could not do till the second. True, he did not get back with a dry skin, but Hansen had had a drenching before, and did not mind it much.

The east wind had worked great changes in the ice; we got up to within three fathoms of its edge and followed it. There were hummocks of considerable height on a shoal three miles from Cape Simpson. As the weather was still very hazy and the gale was stiffening to a hurricane, we sought shelter in the lee of some ground ice close by, and made fast to it. Nothing is more risky than to let the vessel lie near ice which cannot be overlooked; it is easy to get caught in a bay inside a mass of ice without any outlet, and so have a chance of an involuntary trip to the North Pole. And on these borders of the Arctic Ocean the risk is greater than anywhere else. The current that Nansen so splendidly utilised, the current that has carried hundreds of American vessels northwards, and in 1879 carried away the "Jeannette," under De Long, runs strongest near Point Barrow, and sets towards the north-east, sometimes with almost torrential rapidity.

While we were working up against the ice we were unfortunate enough to strike one of the propeller blades against a projecting piece. The engine stopped abruptly, and, notwithstanding all efforts on the part of the engineers, we could not get it to work. The propeller blades were not damaged but the shaft had got warped. We took the matter quietly and soothed our anger by the consoling reflection that it *might* have happened to us very much earlier. The ground ice to which we now made fast was large and safe, and we were protected against any contingency. We were just outside Fatigue Bay. During the night the ice pressed closely; at times it was firmly packed, at others it was loose. The wind lulled, and at 6 A.M. we cast off and proceeded westward. Heavy masses of ice drifted past us and we had to dodge between them. While the mainsail was being set, the gaff broke and we were in a very awkward predicament, no engine and no sail; in fact, without the mainsail the vessel was incapable of manœuvring. We put the trysail up, and we got on pretty well, running before the wind. An hour later the gaff was spliced and the mainsail set again. Now we went smartly ahead to the west among scattered ice, but I did not like seeing so much ice between us and the shore. About noon the ice closed in so much that we made scarely any further progress to speak of. We came to some ice which we thought was frozen to the bottom, and made fast to it. The weather continued foggy. As our moorings appeared to be drifting westwards, we were compelled to get disengaged. At 2 P.M. the fog lifted and we sighted the mastheads of two ships beyond a long low point of the coast. This point must have been Cape Barrow. The vessels were lying to leeward on the west side, waiting for a chance to get ahead. The drift ice separating us from the open channel along the coast began to get looser, and we decided to force our way through. The prospect of getting into communication with the vessels was a further inducement. At 3 p.m. we headed towards a much looser pack of ice; we cast off and with all sail, made straight for the loosest part. I had sent up Lund, the most experienced of us all, into the crow's-nest. Now that we were entirely dependent on our sails, the presence of experienced and tried whale-hunters on board was of inestimable value. To manœuvre a sailing ship in closely packed ice requires many years' experience; anyone can make headway with a steamer. Luckily we had a small stretch of open water in front of us, so that we had got up some speed before we came up to the ice. The "Gjöa" struck it with a heavy thud, turning everything on deck topsy-turvy, but her bow parted the ice. We all worked franctically with boathooks to clear the ice away as well as we could, and the boat pressed forwards under full sail. Thus we forced our way ahead, inch by inch until there was very little left of it. The ship then took a fresh start for the last assault; it seemed as if the old "Gjöa" knew she had reached a critical moment. She had to tackle two large masses of ice that barred her way to the North West Passage; and now she charged again into them to force them asunder and slip through. The lads attacked the ice on both sides with boathooks, a tough desperate fight. The ice yielded a fraction of an inch at a time, but at last it gave way. A wild shout of triumph broke forth when the vessel slipped through. The barrier was broken, we were out in open waters, with a clear homeward track before us.

With joy beaming on our faces we headed westward in the open channel, under full sail. We were now gradually getting into the track of the ships we had sighted, which meant news from home, from our dear ones, and from the outer world. We knew that on board those ships letters awaited us. Bank after bank lay along the coast, forming large lagoons. At 6.30 P.M. a thick misty rain come on, entirely obscuring our view. Again we had to lie up against some ice. An hour later, it got lighter again and we now saw the mastheads of five vessels. It did not take us long to get under sail again, and we were soon well under way. Luckily the wind had kept in the north-east all the time, so that we had not felt the want of our engine. There was plenty of water; the shallowest sounding being two and a-half fathoms. At 10 P.M. we rounded Barrow Point, the north-west point of America. It was late at night, yet we thought we really must hoist our flag, and we did so.

The ships had already sighted us, as a boat was coming towards. It was Einar Mikkelsen, the leader of the Anglo American Polar Expedition. His ship, the "Duchess of Bedford," was lying here waiting for a favourable wind, to take her eastward. We ran close inshore and anchored. What with steam sirens going and flags saluting, there was plenty of stir and excitement. The "Harold Dollars," the tender of the whaling fleet, and the schooner "Monterey" came up to us, and greetings and congratulations poured in. The tender had been lying here a long time and had given up all idea of going any further, which under the circumstances was not to be wondered at. The "Monterey" was one of the two vessels that had managed to slip out through the ice before the autumn. Her crew was now whaling, and did not trouble about eastward regions. A steam launch from the American Revenue cutter "Thetis," lying two miles off, came and brought further visitors. Subsequently I went aboard the tender to inquire after the mail. Whom should I meet there but my old friend, Mogg, my travelling companion in Alaska. He was ice pilot on board. A large parcel of letters and a present of apples and cigars were the welcome reward awaiting me. No one on board the "Gjöa" turned in very early that night. All the letters had to be read and all the news discussed. Every one of us had good news from home.

The first thing we had to attend to now was to procure material for repairing our gaff; the last repair was merely temporary. Next morning I procured four good planks from the tender, and we at once set to work to mend the gaff; it did not take us long. But when it was finished, with all its lashings, it certainly presented a rather patched-up appearance. It was so big that I was almost afraid our winch would not be able to work it. I was on board the tender again at night, and bought from the amiable Captain a quantity of provisions and other things to cheer us during the remainder of our time on board: fruit, cigars, and a quantity of American canned goods, which we highly appreciated. When I got on board the "Gjöa" again, a heavy ice-floe had drifted on to us, and was pressing against our chain. We managed to weigh anchor, set sail, and get clear. Being under sail, we thought we might as well proceed on our voyage at once, and, doubtless to the surprise of the other vessels, we sailed away to the west. The "Treasure" and "Karluk" were just coming in from Herschel Island. At 11 P.M. we passed the "Thetis," making for the other vessels, to maintain order. Of course, we ran up our flag, always pleased to exchange becoming courtesies with other vessels. But the commander of the "Thetis" thought otherwise; the American flag was not hoisted.

The breeze from the north-east still continued, but it was very slack. We had a few squalls with rain. Off Cape Belcher we bade a last good-bye to the ice; it was lying about in small floes. From that time we saw no further trace of it. I was quite prepared to find that we had sprung a leak in our last encounter with the ice; but, on the contrary, the boat was tighter than ever. Instead of two hundred strokes of the pump in one watch, we now needed barely forty. But the old leakage reappeared soon after.

On August 24th we were becalmed for the first time. So long as we had a good wind, it was easy enough to pretend we could do just as well without an engine; but things were different now, and, unable to restrain our annoyance, we vented it on the engineer. "Now then, Smith, why haven't you got the 'coffee-mill' going?" Jeers and jibes poured down through the engine-room skylight on the heads of the unfortunate engineers, who were perspiring in their efforts to repair the defect, if possible.

Since we left the Greenland coast we had never seen a walrus; here we saw several, but never any considerable number of them. The temperature rose daily as we proceeded southward, and we much enjoyed the change. However ardent a Polar explorer one may be, it would be futile to deny that a genial temperature is very agreeable after having been deprived of it for some time. Now and then we saw some sea-fowl—auks and others—emerging from the waves. Even the sight of a jelly-fish was hailed with joy, this being another sign that we had reached milder regions.

On August 30th, at 11 A.M., we sighted Cape Prince of Wales. This marks the eastern entrance to Behring Strait. But, as the summit was shrouded in fog, we could not be sure whether this was the point. From this cape a long, narrow sand-bank extends twenty-five miles to the northward. There is deep water on both sides, so that it is not easy to determine by soundings on which side of the vessel the sand-bank is situated. Unless we were lucky, we might easily get between the shoal and the shore, and in that case our position might become rather serious under existing conditions. There was a high sea running, with a succession of squalls, and the "Gjöa" pitched heavily. To be on the safe side we steered for the open sea as soon as we sighted land. At 1.30 P.M. the weather cleared, and we sighted Fairway Rock, a peculiar hayrick-shaped rock, rising sheer out of the sea. We could not have had a better landmark, and we were now able to get into our course again. When we got down into the Strait, we caught a slight glimpse of the Diomedes Islands. These look barren and inhospitable, yet they are the abode of a whole Eskimo tribe, who are all, both men and women, very popular with the whalers. There is no harbour there, but when going north the whalers always call in to barter for various supplies. They are also glad to engage Eskimo from there, as they are considered able hands.

ANTON LUND (SPRING, 1906).

As we passed between the islands and the shore, we "old hands" gathered on deck and drank the first cup

Godfred Hansen. Anton Lund.
Roald Amundsen. Peder Ristvedt. Adolf Lindström. Helmer Hansen.

AFTER OUR ARRIVAL AT NOME.

SPORTS AT NOME.

to celebrate the final accomplishment of the North West Passage by ship. I had hoped to have a little festivity to mark this notable event, but weather did not permit. The event was celebrated by a simple toast, nothing more. We could not even hoist the flag, as it would have been quickly blown to tatters.

I had intended to reach Cape York before it was dark, and lie there for the night, but we did not get so far.

We lay about ten miles off Cape York with double-reefed mainsail, reefed forestaysail and standing jib. Cape Prince of Wales sheltered us to a great extent from the seas. When we shook out the reefs at daybreak our gaff broke. This time it was the great weight of the gaff itself which was the main cause of the catastrophe. Our only hope now was that the wind would hold out till we reached Nome, and could procure a new gaff. The old one was beyond repair. We headed south towards Nome under trysail and any other sails we could carry. Originally, I had no intention of calling at Nome, but after the accident with the gaff we had no choice. I had received a letter at Point Barrow inviting us to call in on our way south and accept the hospitality of the town of Nome. This suited us well now. We took the shortest route, passing eastward along the Nome coast: during the afternoon the wind gradually lulled down to a faint breeze, and we made very little headway.

"Well, Lund," I said, as I was sauntering on deck that beautiful afternoon, "you are equal to any emergency, can't you get the mainsail set?" This was too much for Lund's pride, and soon after that the mainsail was up. The "Gjöa" did not look quite as trim as a pleasure yacht, but, at any rate, we travelled a good deal faster, and ran ahead smartly before the light breeze. The houses in Nome were soon visible on the horizon; if the breeze would only last another hour, we should be there. But fate had decreed it otherwise, and a dead calm set in. They ought to be able to see us from the town now, so we hoisted our flag. An occasional slight puff of wind carried us a little way ahead, but it did not amount to much. As daylight faded, we saw that the lamps were being lighted in the town, and our position was a very tedious one. Suddenly a steam launch appeared in front of us, and we heard whistling, shouting, and cheering,—the American's mode of expressing enthusiasm. Dark as it was, we could still discern the Norwegian flag floating side by side with the Stars and Stripes on the launch. So we had been recognised.

The reception they gave us at Nome defies my powers of description. The heartiness with which we were welcomed, the unbounded enthusiasm of which the "Gjöa" was the object, will always remain one of my brightest memories of our return. Nome has no harbour; it lies on the open coast. We were, therefore, obliged to anchor well off the shore, and keep ready to weigh anchor as soon as the wind rose. After we had anchored, Lieutenant Hansen and I went aboard the launch to greet our amiable hosts and hostesses, and we were taken ashore. An electric searchlight on the shore played on

THE "GJÖA" AT ANCHOR OFF NOME.

the boat all the time. As we approached the shore, we were so dazzled by the powerful light shining right into our eyes that we could not see anything. The boat touched land. I really cannot say how I got ashore, but a jubilant roar of welcome issued from a thousand throats, and through the darkness of the night a sound burst forth that thrilled me through and through, bringing tears to my eyes; it was the strains of our national air—

"Ja vi elsker dette landet."
(Yea, we cherish this our country.)

SUPPLEMENT.

TOWARDS KING HAAKON VII'S LAND.

By First Lieutenant GODFRED HANSEN, Vice-Commander of the Expedition.

Introduction.

WHILE sailing through the narrow, shallow straits of the North West Passage we were all fully occupied with the navigation of the ship in the strictest acceptance of the word, consequently the cartographical work undertaken during the "Gjöa" Expedition was carried out on sledge and boat trips, some of which it fell to my lot to lead.

The most important were :—

1. A sledge trip to Point Richardson on the American mainland in March, 1904, on which we discovered two islands in Simpson Strait which were called after Commander A. P. Hovgaard. The trip lasted ten days, and my companion, Ristvedt, and I had to draw the sledge ourselves, because the dogs which had been spared by the dog sickness were away on a sledge trip with Captain Amundsen.

2. A boat trip westwards to Cape Crozier. The object was to investigate the conditions in the narrow

parts of Simpson Strait with reference to the passage of the "Gjöa" during the next summer, and at the same time to take a depôt of pemmican and dog's food of about 500 lbs. out to Cape Crozier. The trip lasted from August 6th to the middle of September, 1904. Hansen accompanied me and sailed the boat.

On the way we found some skulls and bones of two white men. They were lying scattered over the low foreshore at Point C. F. Hall, and had been placed by this Arctic explorer under a stone cairn. Close by we found the stone on which he had cut the words "Eternal Honour to the Discoverers of the North-West Passage." We collected the bones together again and covered them over with stones, on the top of which we placed Hall's stone.

3. A sledge trip in the spring of 1905 to Victoria Land, and along its unknown east coast.

In what follows I shall endeavour to describe this trip more in detail. Ristvedt accompanied me. I owe a good deal to his equanimity, his constant good humour, his indomitable energy, as well as to his excellent shooting.

Before starting I drew up the following report for Captain Amundsen :—

To the Chief of the Norwegian "Gjöa" Expedition.

I beg to inform you of the preparations for the sledge trip ordered by you for charting the unknown western stretch of coast along M'Clintock Channel.

The Expedition consists of two men with two sledges and twelve dogs.

The stores for the men have been determined mainly in accordance with your normal list, and are as follows :—

Normal list for daily consumption per man—

	Grammes.
Margarine	30
Chocolate	200
Bread	300
Pemmican	200
Green stuffs	25
Pea flour	25
Dried bilberries	5
Sugar	5
Coffee	5
Salt and pepper	5
Total	800

The list of provisions for two men for seventy days will therefore be as follows :—

	Kilos.
Margarine	4
Chocolate	28
Bread	42
Pemmican	28
Green stuffs	3½
Pea flour	3½
Dried bilberries	1
Sugar	1
Coffee	1
Salt and pepper	1
Total	113

The depôt established in the summer of 1904 on Cape Crozier consists of:—

100 kilos. of fish and suet melted together, and
130 kilos. of pemmican for dog's food.

If the journey from Gjöahavn, Ogchoktu, to Cape Crozier takes seven days, and a depôt of stores sufficient for seven days is left there for the return journey, we get supplies sufficient for fifty-five days.

If the Cape Crozier depôt is destroyed, then we can only continue for thirty-four days from that point, but in that case it will be necessary to kill four dogs.

The itinerary is as follows:—

The Expedition will be ready to start on April 1st. We shall work along Simpson Strait to Cape Crozier where a report will be deposited; thence the course will be due east towards the highest island in the group marked "land seen by Rae." This group of islands will be surveyed and the course shaped due north over Driftwood Point, Cape Alfred and Pelly Point, to Collinson's Farthest, and from thence out into the unknown towards Glenelg Bay. The distance as the crow flies is about 850 miles, so that the Expedition may be expected back in Gjöahavn at the beginning of June.

If, on the return trip, Victoria Strait should be impassable, or should there be any other obstacle to prevent the Expedition reaching Ogchoktu by July 15th at the latest, we shall work down towards Cape Colborne (Dease Strait) which is said by Collinson to be low and sandy. A cairn will then be erected at some visible spot.

Then followed a review of the survey work carried out up to that time, as it was by no means so very certain that we should ever return.

Our sledges were ready on April 1st. Ristvedt's loaded with 500 lbs., mine with 450 lbs. I was to drive in front, as we thought it would be easier in that way to keep the rear dogs up to the scratch. We had iced the runners in the Eskimo fashion. But what with wind and snow the weather was too bad to start, so we postponed our departure till the next morning.

April 2nd broke with good travelling weather. It was still blowing fresh and the atmosphere was very hazy, but the temperature was only down to about 25° Fahr. so that the little wind there was would only act on us hardy Arctic navigators like a breath of summer from the distant south. When the great feat of the day in these latitudes, indeed in any latitude I have ever been in, namely, getting up in the morning, was accomplished, I went up and looked out at the weather, and found it suitable enough to make a start. Then when Ristvedt came down, I said: "Well, so we are off to-day, eh?" Yes, as far as he was concerned there was nothing to prevent us starting at once. Of course, we knew we were leaving the flesh-pots, the warm bunks, the fire-side and bright lamp; but now the winter was over we wanted some fresh air in our lungs, and some under our wings too. Both of us loved Nature, and we wanted to see

LIEUTENANT HANSEN (SPRING, 1906).

what we were made of when we should meet her out in her own kingdom, because she is not only seductively attractive, she is also cutting and severe. We were anxious to see if cold would dull our brains, want curb our energies, monotony deaden our senses, or whether, as we imagined, we should be the masters; whether in fact, there was real manhood behind our own self-confidence. Now we were ready to start: the sledges were equipped as well as could be with our previous experience and with the aid of all the resources placed at our disposal by the main Expedition.

When you are ready to start, when there is absolutely nothing further to do than to sit with your arms folded, if the departure has to be postponed, you are apt to get very impatient and restless at the delay. You keep getting up to look at the weather. If it is a little better, you ask yourself if you ought not to start; the very human desire not to be outdone by others makes you think it strange that you don't make a start. Such a state of mind is by no means pleasant, so now that we had decided to go, it was like a load taken off our minds.

Lindström's excellent coffee and reindeer steak were very tasty. We had had cake on the previous day, and Lindström was too sensible to give us cake again, "No," said he, "you have already had your farewell cake once, and you must make it do. Someone might always be making a pretence of starting next day." When breakfast was over we went out to harness up. Hansen and Wiik fetched the dogs and put them to the sledges.

We two travellers were to be spared as much trouble as possible. Starting on such a protracted trip is not altogether a festive occasion. Your comrades are anxious to express their best wishes for the journey and a happy return. They know what you are going to encounter, for they have all had a turn at it, and they show their goodwill up to the last moment by taking all the work on themselves, even that of starting the sledges. Unfortunately the ice fell off the runners as they loosened the sledges, the mild weather having rendered the ice covering too soft, so that it was torn off just as the sledges were twisted aside. This is called freeing the sledges. If the sledges stand still for a long time the under part of the runners freezes fast to the snow, so that the sledge driver has to turn the hind part of the sledge to one side or the other. It is very seldom possible to get the dogs to do this, it being often too heavy for them, and they won't put their shoulders to the wheel till they see that the sledge is free.

We were now clear for starting. The sledge was put in commission; that is to say, I fastened my Norwegian sledge flag at the back of the sledge over the cyclometer. Then I arranged the photographic apparatus, that those we were leaving could obtain a good snapshot of us. Yet another hand-shake all round, the last "good luck" from the Captain, and off we went with the dogs at a comfortable trot towards Fram Point. The first halt was made when we passed Fram Point and turned round behind Fram Hill, so that we could not be seen from the ship. While we were resting at Fram Point we saw Talurnakto come running as hard as his short legs would

PEDER RISTVEDT (SPRING, 1906).

carry him. He came up with an igloo knife, a parting gift from the Captain to Ristvedt. Igloo knives were made of the blades of some large carving knives, and were fastened to long, flat, wooden handles made by Hansen. We had only a few of them, and they were highly prized by the Eskimo, so it was necessary to keep them in reserve in case an opportunity occurred to buy any desirable object for the ethnological collection of the Expedition. We had had one each given to us, but Ristvedt had unfortunately lost his ; but now he had one presented to him for use on this trip, and Talurnakto went back with thanks and kind messages. Then a sharp pull on the sledges, a call to the dogs, and we started again along Petersen's Bay, over towards Snadde Hill.

When we reached this spot, Ristvedt's dogs were already fatigued, but it seemed to me too early to finish up the day's travelling. We had intended to pass the night at Svarteklid, in the igloo we had built on our observation journey in the month of February. We then drove down on the ice again, following the coast until we turned in over the land towards the northerly Kaorka Isthmus. On the top of this we turned round, and sighted the mast of the "Gjöa," like a plain black line on a grey background, grey clouds in the sky, grey haze in the atmosphere, so dimming the light that even the snow looked grey. With the glasses we could still see the flag waving from her top mast. Then we went further on, down the slope towards Kaorka Lake, and south along Black Hill slope, and it was not long before the dogs got scent of the igloos. This, I am thankful to say, put some life into them, which was welcome, as we were rather tired of whipping and shouting, as we had been compelled to do, the whole day long; but this renewed vigour had the sad result of overturning me twice with my load, and, as 500 lbs. are no small weight to lift, it was no wonder it made my back ache. The first time the sledge turned over, I was able to right it again by myself, but the second time Ristvedt had to help me; and our trials and troubles for that day came to an end five minutes after we drove up in front of the igloo and stopped after a day's march of nine and a-half miles. My dogs were fresh and ready for new efforts, whilst Ristvedt's team was almost done up, but we hoped that they would be better with a little practice. When we stopped I made a pretty little speech. All we could afford was a little drop of rum, and in this we had to toast all that should be toasted on such an occasion. We drank to the man who sent us out on the trip, to our good comradeship in the times to come, to reaching our goal near "Wynniat's (Collinson's) Farthest," to a happy return full of honour to the flag under which we travelled, and then the rum slid down our throats. Should anyone now ask me if I enjoyed it, I could confidently reply "Yes." Sledge driving on long journeys is not what one sees in pictures of Greenland, sitting in warm furs, and cracking the whip as the dogs fly like race horses. There is no such racing with the heavily loaded sledges necessary for such a long trip as ours. It is only on exceptionally good ice that one can even sit on the sledge at all, and at first one often has to be the beast of burden, especially on stretches of land covered with soft snow or rugged uneven ice.

Although we had only covered nine and a-half miles, not a very long distance in itself, yet we were thoroughly tired out and hungry. The coffee in the morning and Lindström's reindeer steak was all we had had the whole day, so that our hunger was easily explained. Then we had the rum. We saw its gold-brown sheen in the silvery aluminium cup; our hands lifted the cup, and we sniffed the spicy fragrance like a breath from a sunburnt southern plantation wafted over the barren ice plains. The cup reaches the mouth safely and surely, for an Arctic explorer's hand never shakes, and so it slides down, ice cold, refreshing, heating, and invigorating. You may turn up your teetotal noses, but I know how useful alcohol is at such a time. Let me tell you one thing; I have known what it is to get up healthy and fresh in the morning and drive out with my sledge; I have expanded my chest and felt the fresh air filling my lungs, felt the blood circulating in my veins, felt as if I had strength enough to run to the end of the world; the beauty of the sun and sky, all Nature filled my soul with the most delightful sensation; but when evening came I was a wreck, I had lost faith in my luck, I was dead tired, fagged in every limb, the brain benumbed, my only desire being to keep on till I dropped; I shuddered at the thought of the effort to break loose from the monotonous toil in the traces, to pitch the camp, although I knew from experience that the sleeping-bag was the remedy for all this. It is in such conditions I have found alcohol to be useful; and, even when the tiredness and exhaustion are not quite so utter, the stimulating action on an exhausted man with an empty stomach is astonishing. The tent is pitched, the dogs are fed, and the food is boiling in half the time it would otherwise take; one evades much of the effect of piercing wind and cold inseparably connected with the pitching of a camp when heated from the day's toil. When the reaction of the alcohol sets in you have been in your sleeping-bag some time, so that this reaction has almost the advantage of obviating the difficulty in getting to sleep owing to over-fatigue.

On this particular occasion the pick-me-up had also the good result of making us conduct ourselves like Christians, when we found that the gentleman who had used the igloo last had neglected to close it properly, so that a mass of snow had driven in. Under other circumstances our expressions would hardly have been so forbearing; now we simply made a hole in the wall as the entrance was quite impassable, and some snow had penetrated the house itself; but there was still sufficient room for a couple of sleeping bags. Then we got ready for the night, fed the dogs, crept into the house, and sealed up the entrance with blocks of snow. This we did with very great care, for experience had taught us what a quantity of cold can get in through a crevice not larger than a keyhole. The possibility of living in an igloo depends altogether on its being air-tight.

On April 4th we succeeded in covering a greater distance, 11 miles. Ristvedt, from his military training, had a theory that in field service it is always the second day's march that one feels the most, and this, indeed, seems to be correct, for the third day everything went much easier. We passed Point C. F. Hall, where some of Franklin's men lie buried, with our flag hoisted; but we did not use the flag more than necessary, as the wind and weather would soon have turned it into a ragged trophy. It was also very inconvenient to carry it hoisted, as I always got the lash of my whip wound round the staff just at a critical moment; but here we had to wave the flag, and this we repeated every time we passed the cairn. We never omitted this. Over this lonesome grave, on a stony headland in the remotest seas, a spirit of solemnity seems to hover. Once you have seen the stony beach with its little sea, the creek just below, and the low ridge of King William Land stretching beyond, losing itself in the greyish haze of the winter's light, you can never forget the sight. Bitterly sad was the lot of these two men, and that alone would constrain us to wave a respectful salute, though we also desired to honour the memory of their deeds.

April 14th was the exciting day when we were to arrive at the depôt, and discuss whether our further progress was assured. It was very beautiful weather when we started. The ice had now quite a different appearance. Out over the Straits it was broken up into a mass of small packs, but where we were travelling it was in fairly good condition, in a small belt along the land. Soon after starting, Ristvedt went on land in pursuit of a flock of ptarmigan sitting on the crest of a hill. They disappointed him by taking to flight; but we were soon compensated for this, for, while I was driving on, and had already got a good way ahead of Ristvedt, I suddenly saw something a long way off, which looked like a stone. If it was a stone it must have been a very large one, judging from the distance, and I could not remember having seen any exceptionally large stone at this spot at the time of the boat trip. So I took out the glasses, and there, sure enough, was a reindeer. It can be well imagined that I immediately stopped the dogs, lest they should scent the quarry and spoil the sport. I waited quietly till Ristvedt came up, and then took charge of both the sledges. Ristvedt had long proved himself a much better hunter than I was, and, however pleasant it might have been for me to have a shot, this was not the time for practising. We had to get the best possible result from our cartridges, no misses being allowed, and, therefore, Ristvedt had to go out whenever anything living appeared on the horizon. Someone must remain with the sledges, otherwise it would have been impossible to keep the dogs quiet. It was hard enough to manage this with only myself in charge, especially later on, when the dogs knew what Ristvedt's absence with his gun meant.

The reindeer stood out on the flat plain, it was impossible to get any cover, and one could not help being seen when advancing like a black spot over a white surface. But Ristvedt adopted the Eskimo method: making a long detour, until he had got the sun at his back, he made straight for the deer with his head bent down, so that it did not project above his shoulders, and only moved his legs from the knees downwards. This is a useful method, as the Eskimo secure many deer with it, although they must approach near enough to be able to use their bows and arrows. Ristvedt only wanted to get within a couple of hundred yards. He was a sure shot at that range. It was very exciting to follow the hunt. At the moment there was hardly any wind, and, what there was, was in such a direction that the deer did not get any scent. When Ristvedt had proceeded a certain distance in this fatiguing crouching position, I saw the deer lift his head and look towards him. It was evidently speculating as to what it could be. Ristvedt's height was about the same as its own, and the breadth about the same as that of a deer when approaching direct. The sun was straight in its eyes, and made it blink; but there was surely no danger, it must only be a comrade. It then lay down, apparently thinking: "So I can go on digging in the snow and I must not lose any time if I am to get a meal to-day." This was its last thought; looking through the glasses I saw it fall as if struck by lightning, and then the short, sharp report of the gun reached my ear. The dogs started up with stretched-out necks, pointed ears, and distended nostrils. A crack with the whip and we started at a dashing rate towards the spot from whence the report had come. I hardly had time to jump on the sledge before we were off. When we reached Ristvedt, who stood by the side of the deer, the dogs stopped of themselves and began to lick up the snow where the blood had run, and I had to use the whip to get them to lie still until we could get the deer on to the sledge. We then scanned the horizon with the glasses, and perceived another deer; off went Ristvedt again, and we bagged a second.

This was a nice load to drive, two newly killed deer: splendid food for us as well as for our dogs. We had not far to go to the depôt at Cape Crozier, so we preferred to load the deer as they were on the sledge and drive off with them. If we had opened them first we would have risked getting much blood over our sledges; we were, moreover, so anxious to see if the depôt was in order that we did not care to pitch camp on the spot, although we had by this extra store of provisions added, so to speak, to the days available. We then drove off across Low Water Creek, a bay direct south from the Cape depôt, and so called because the entrance was so low that, at the time of our summer trip, we could not get our boat into it. Suddenly my dogs scented something. First Silla raised her head in the air and sniffed, but then settled quietly down in the harness, so that I thought I had made a mistake. But soon she began to get restless again. Mylius and Gjöa also began to lift their heads. It was better to stop and see if their noses were better than my eyes. It seemed to me, however, that I could see everything clear right away to the ridge of Cape Crozier, forming a white, sharp line against the sky, a few miles off. Armed with the glasses I saw that the dogs were right; far away up on the crest of the ridge there were two deer, walking peacefully. A slight breeze was blowing down from them towards us. Ristvedt overtook me. Some of his dogs had also been a little restless, and we set our course straight towards the deer. We went along at a sharp trot over the snow, and at a suitable distance we stopped again, and Ristvedt went on alone. He had soon to lie flat on his stomach; he could not have gone straight up the hill towards the deer, which had a bird's-eye view of him. But the deer, which had probably seen the sledges in the distance, were very curious and wanted to come a little nearer to find out what the black thing could be. I lay on one load with the glasses to my eyes. When looking thus, it is difficult to imagine why the man does not shoot, because through the glasses, the distance between the huntsman and the animal seems so short. At last there is a report, and there lay one animal; the other rushed quickly back, stopped, probably wondering why his mate had lain down; then he approached again. If his comrade could lie so quietly, surely there could be no danger. Step by step he came nearer, with his head raised so that the antlers lay back over his neck, stopped, drew a little back, stopped, — off went the gun. The animal wheeled around and dashed off at full speed. "He has missed him" thought I. But one can't run away from death.

Bay. Silla. Per.

OUR FAITHFUL COMPANIONS.

Death had gone out from Ristvedt's gun and it was now draining the heart's blood out of the deer on to the white snow Fifty yards—quicker and quicker—one hundred yards—a heavy fall, and all was over. "No, he hit him after all," thought I. I let Silla loose so that she could follow the deer in case he should rise again, but there was no need. I then drove up to the hunter with the sledges and we placed the two fresh deer on them. Now with really heavy loads we continued along the ridge; we should soon be there, surely, as it was just around that projection. Quite right, there lay the cairn large and broad on the foreshore; and the depôt—the bears had robbed it.

We pitched our tent, skinned the deer, and went to bed. But that night, for the first time, I slept badly; as the four meagre reindeer by no means made up for my ample depôt, how could I now hope to cover a little new ground? New plans ran unceasingly through my head. Supposing the dogs should break loose and eat the meat lying outside unprotected on the snow. Every movement made me listen attentively. Yes, and the bilberry-rum punch we had drunk as a cup of rejoicing for the unexpected deer, and a cup of sorrow for the loss of the depôt, would have been better left alone.

Next day, April 15th, we had brilliant sunny weather, which we utilised for drying our wet skin clothes. They get wet at this time of the year, partly from inside and partly from outside, from the snowy mist, which settles on you and thaws when the sun shines and remains matted in your dark clothes, one of the few dark spots in the landscape. We divided up the reindeer meat and had to re-arrange our loads. We had now dog's meat for thirty days added to what we had already. This permitted of an advance of twenty days, if we were prepared to sacrifice some of our dogs for the advantage of their comrades, should luck desert us, and should we find no game on the way. We established a depôt of deer's meat on the hill for the dogs and for ourselves for four days. We put it amongst some loose stones on an old shore line. By pulling the stones aside we made a large hole in the earth. Alongside we placed a tin containing two and a half gallons of petroleum which we thought we would dispense with if we could only continue for about twenty days more, then we rolled large stones on top of it. Foxes could not touch it, and we hoped that we had placed it so far on to the land that the bears could not get a scent of it. As a rule bears do not care to leave the ice. We, therefore, hoped to be able to find the depôt safe again, otherwise I really do not know how property can be protected against bears. The depôt at Cape Crozier, which was established at the time of the boat trip, consisted, as previously stated, of 500 lbs., partly of pemmican, partly of a mixture of fish and suet, for dog's food, and put into two soldered metal cases. At Cape Crozier there are huge rocks which disintegrate into large flat slabs. We had set the two tin cases on the foreshore and arranged round them regular vaults made of heavy stones, which two strong men could hardly carry. We had thus passed half the day in piling stone blocks over the whole depôt, stones as large as any of us were capable of lifting. It was quite a little mountain when it was ready, but all this the bears had torn aside, and all we found was a single piece of metal rolled together. The bear had set his mark on it—five long rents through it lengthwise. He had thus ripped the case open, but why he had afterwards rolled the metal together and bitten it I am unable to say; I trust it was from rage at having cut himself with it.

Although we had taken thirteen days to reach Cape Crozier (I had calculated on seven) I was sanguine enough to think that we could drive home in five. It was, as a matter of fact, only one hundred miles, and at a better time of the year and with lighter dogs it ought to be easily done, even if we were forced to kill some of the dogs; therefore I left behind stores for only four days. For lunch that day we had marrow soup, made of the marrow bones of the four deer. One ought not to be too greedy when eating such a dish; the rich hot fatty stuff glides down so easily when one is hungry after a tiring day. But it slipped down on too large a scale. Ristvedt, who had a stomach like a harmonium, although he was not a sailor, maliciously told me afterwards, when the meal was well down, that on the occasion of the sledge trip in the previous year he had treated the Captain to a similar dish, with the result that he had a dreadful stomach ache; and, as a matter of fact, I myself did not escape it.

On Sunday, April 16th, we stopped on account of bad weather. We only took a little trip inland and saw that inside the ridge at Cape Crozier there is a depression running right across the land from Alexander Strait to Simpson Strait. Two ptarmigan flew by us and there seemed to be no scarcity of them. We had already seen some in fact, so that there was every prospect of getting a few into our stew pot if we cared for that sort of sport. We also saw the first snow bunting; it flew in front of us, alighting here and there, pecking at the snow. Where the ridge projected bastion-like towards the depôt point, we built a cairn. I hoped to be able to see this from "Land seen by Rae," so as to have a definite point to refer to. We then stole down and settled into our sleeping bags; we were frozen, as the wind was blowing hard. The snow drifted in and the temperature was nearly 4° below zero Fahr., but the snow-bunting sat chirping on the top of the cairn.

The pack ice lay waiting for us. We had seen the surface of the ice over the straits slowly changing its level appearance ever since we left Fitz-James Island, but it was only now as we looked from the ridge of land on Cape Crozier towards the "Land seen by Rae" that we failed to see level surface anywhere. Nevertheless, we slept comfortably that night, for we did not yet know what pack ice was; we had only heard of it and had been told that one had to struggle and wriggle through it, and that as one advanced it flowed with the current just as rapidly in the opposite direction. We knew also that those who had told us of such difficulties were not easily frightened, but we comforted ourselves with the thought that the ice we had to pass over was only "Strait ice." There were no mountain-high ridges of pack ice to be seen there. It could only be trifling in comparison with what may be met with in the great Arctic Sea; besides it was quite still, the frost had solidly united it from coast to coast in this narrow passage, so we did not risk drifting out of our way. We should probably not be able to do more than ten miles a day, but as it could not be much more than about fifty miles to Victoria Land, we should there have "Land seen by Rae" to travel on. This was consolation enough and the last night on Cape Crozier we slept very peacefully. The next evening we were not so hopeful; in fact, there lay the pack ice still waiting for us, and after travelling on it one day, we knew what we were to expect of it.

The first day, April 17th, we travelled from 9 till 3 o'clock; I could do no more. During this time, we had been travelling at the magnificent rate of half a mile an hour, that is to say, we had covered three miles in all. We did not talk much together that evening, we were too tired. However, we agreed that we could not go on like this and that we should have to find some other method of progression if there was much more of that kind of ice. The next day we endeavoured to drive on with one sledge at a time, harnessing all the dogs to it, but this made the way three times as long, because we had to return for the other sledge.

On April 19th we started by going some distance on

ski and then turning back to fetch the sledges. In this way the dogs followed the track made, and we could devote all our attention to steadying the sledge so that there were comparatively few upsets. But it was very laborious work struggling on, half crippled. It was a brilliantly clear day about 13° below freezing point; in front of us the land lay so near that it had quite lost its monotonous white tone, and we could already see all the details of the landscape. After three hours' more work we got out of the pack-ice; our toil was over for the present, for, after all we had gone through, it was only child's play driving over the level strip of ice that separated us from the land. In a couple of hours we reached the shore, but to our great astonishment we had first to drive over two islets, some sixty feet high before we made the land from whose shore rises the very conspicuous Mount Rae to a height of about 330 feet above sea-level. In the hazy light it was impossible to see both islands; it all appeared to form one unbroken surface sloping up to the summit.

The day after, April 20th, we set our course towards the land we had seen out westwards from the top of Mount Rae. The route led us past very fresh bear tracks, but we never caught sight of the bear; while we were resting out on a little islet in the middle of the sound, a ptarmigan flew up and sat down close by Ristvedt's sledge, but it had not sat there long before it was shot. It sat long enough, however, to lead me to think how strange it is that ptarmigan are so shy at times that it is almost impossible to get within range, at other times are so tame that they come and sit cackling beside you and remain there even if you have to fidget about to get hold of your gun.

When we reached the other side of the sound, to the large low island we afterwards named "Easter Island," we saw some reindeer. As the dogs had not noticed anything, we thought they would be quiet enough if we

Godfred Hansen.
ON THE WAY TO VICTORIA LAND.

tied them so that they could not start off after us with the sledges. Then each of us went out in a different direction, but neither of us had any luck. Ristvedt was the first to scare his quarry, and the deer came at full speed down towards where I was lying; I crouched down as much as possible, but it kept out of range. I saw Ristvedt going to the sledges, and, as the two deer I had gone after were still at some distance, I waited quietly until

Ristvedt came up. Then he went after the deer. They must, however, have seen the other run away, as they were so frightened that they started off long before he got near them. Then I drew up after Ristvedt, and on the way I passed a deer track, which I thought was made by the deer Ristvedt had gone after. I therefore let the dogs follow up the fresh track as fast as they liked, as I thought it would certainly lead me to Ristvedt; but in this I was mistaken, for this was the track of the first deer that had been frightened away. The deer had long ago disappeared in the distance, but these tracks were fresh, and we followed them at such a speed that I could neither turn the dogs aside nor get them to stop; they had got the thirst for blood. They started with their noses down in the snow, like a couple of famished wolves. Ristvedt got smaller and smaller in the distance, and soon I could see nothing but the white snow-field all round. Ristvedt's sledge had stopped long ago. His dogs were exhausted, and this was fortunate, under the circumstances. At last I succeeded in turning my sledge over; this stopped the dogs, and it was now my turn. After giving them a good drubbing, I turned them round towards Ristvedt. I then righted the sledge and started back. I ran alongside the sledge, so as to be ready to turn it over, if necessary, but they had caught sight of Ristvedt and there was no further trouble. Ristvedt was standing wondering where I was off to. We then started again in the usual order. We shaped our course towards the setting sun, which glided like a glowing ball along the low ridge in the west. The ridge turned towards us its coal-black, shady side, wherein every detail was lost in the thick darkness. But the crest formed a sharp, irregular, notched line of rocks against the red glow of the evening sky. Suddenly a deer was seen standing right in the sun, its silhouette sharply outlined against the sky. It must have been lying down, and had now got restless. It really looked gigantic as it stood there, like some mighty creature from prehistoric times. For a moment it stood with its head erect, the neck curved back so that the antlers spread over its back. The dogs saw it, but did not seem to realise what this colossal form was; yet they stood, suddenly riveted to the spot, with their legs extended. Men and dogs were bathed in the rays of the sun. The reindeer showed no sign of fear. He simply turned round, and slowly disappeared behind the ridge. Then the dogs seemed to realise that it was a deer, and wanted to rush forward. I had to throw myself on Per and Bay to force them to the ground. Silla and Gjöa were both mad for the chase, so I seized one under each arm, and, rolling over them, held them all in a struggling mass, so as to prevent them from howling. Ristvedt came up; his dogs had not seen anything, and they were quieter. I told him to go on. I hardly expected that the deer would still be walking on the other side of the ridge, for the few howls which had escaped from the throats of Bay and Gjöa were enough to frighten an animal whose race has for centuries been hunted, terror-stricken at the ear-piercing, diabolical howl of the wolves, the solitary sound that wakes the majestic stillness of the winter's night.

Ristvedt went on and reached the top of the hill; then I saw him lie down and rest his gun on a boulder, so I knew that the deer had not long to live. The gun went off and I loosed the dogs. I had just time to get out of the way of the sledge as it dashed past me like a gust of wind. One would hardly have thought that ten minutes ago the dogs were exhausted, with their tongues hanging from their mouths; but now, having scented blood, no effort was too great for them. We had not far to go; only over the crest of the hill. When we reached the other side, Ristvedt was skinning the reindeer.

On Easter Eve, April 22nd, there was a southerly gale and we had to remain in the tent. Eventually, on the Monday, April 24th, there was nothing to prevent us from proceeding. It was necessary to advance, and at first it was easy travelling. The mile down to the coast was a mere nothing, and the next mile and a-half out presented no particular difficulty, but the route gradually became more trying, and we were at last brought to a standstill. I had to go and search for a path, and after forging ahead for some two hundred yards, I reached a level track—but what a track! It looked level enough between the bends, but we continually sank in it up to our armpits. Indeed, it took us two hours to take the sledges over the two hundred odd yards.

Thursday, May 4th, was a great day for the Expedition. First we drove as we did in Simpson Strait, in what we called the good old days, each man alongside his sledge, and sometimes on it; but I had very soon to go on in front, as it was by no means so level towards the land as it looked at first sight. There were hummocks and pack ice here and there in the ice fields and to find the best course, one had to go on in front. The dogs seemed to have a most astonishing inclination to go north-west, and not straight out in the direction we wanted them to take, towards the nearest land. When we came to an extra large hummock I wanted to go up to inspect the route. Hardly had I stopped when Ristveldt called to me that he saw something dark which was moving on the ice a long distance away. What could it be? I almost felt my heart beating. It mattered little what it was; any living creature out here in the barren ice fields offered prospects of further progress. I took out the glasses and found it was not one dark spot; there were many spots spread out in the form of a crescent over the ice. They were Eskimo, seal fishing. So it was by no means strange that the dogs had wanted to go in that direction. We fixed the flag on the sledge and as we thought there was a slight possibility that there might be white folk among them, we began to rub ourselves over with a little snow, to remove from our faces as much of the soot from the petroleum stove as possible. Then we drove on again. Evidently the Eskimo had also seen us, for they came rushing over towards us, closing up as they approached. When they were all collected together at about 400 to 500 yards from us, they halted, and we did the same. Out in these wild regions, where might is right, one never knows whether one is meeting friend or foe, so that it is best to be prepared; the usual formalities of an Arctic meeting had to be complied with. We drove the sledges up so that they were broadside on towards the strangers; Ristvedt lay down behind them with his gun loaded and the cartridges ready at his side. Then I braced up my courage and went forward with my hands raised above my head to show that I had no weapon; an envoy also came out from the Eskimo, holding his hands in the air so that there was nothing to fear, and we met in the middle of the arena. His white teeth shone in front of me, so broad was the smile that spread over his flat amiable Eskimo features. He was not afraid of me and there was no expression of fear in his eyes at the sight of my strange features; indeed, the dirt which covered my face was of the same kind as on his own. I returned his smile with genuine pleasure. Such a meeting at least indicated as much food as we could carry on our sledges. When we came up to each other he said something about "Kilnermium Innuit," the name of his tribe. I understood that, because I already knew the name, and I replied that we were "Kabluna" or white men. Then we embraced and rubbed our cheeks together. When you are in Rome you must do as Rome does. He was my friend for the two days we remained there, and during that time he certainly thought I understood everything he said, merely because I had said we were "Kabluna" when he mentioned the name of his race, but of course I did not understand a word. As our Norwegian-Eskimo language was of no use to us, we could not get any information about the land further ahead, and any conversation which had a definite object, had to be carried on by signs.

When our two parties, that is to say, Ristvedt and the other Eskimo, saw that affairs had taken a peaceful turn, they also approached the spot where we both stood, chatting and gesticulating eagerly. With joy and gladness we all rubbed cheeks together, and then started off to their igloo settlement, a couple of miles away. We went pretty quickly, for they attached what dogs they had to the front of our sledges. When we approached the igloos more people came out to meet us, which led to more rubbing of the cheeks, so that when it was all over our cheeks were almost clean. We would not move into their igloos, as at that time of the year the snow house is a very miserable dwelling; the roof melts with the heat which quickly forms inside, owing to the temperature of the outer atmosphere, so that we preferred to remain in the tent which we raised, surrounded by a large crowd of Eskimo filled with curiosity. The spectators were highly amused, but fortunately we had Lili to keep them away from the sledges, so that we avoided any unpleasant results from their curiosity.

There were about half a score of igloos, although some

of them were empty, as several of the tribe had gone northwards towards Admiralty Island. There were about twenty to thirty individuals in this camp, and as far as we could understand, about a similar number had gone north; but, on the whole, they were only a very small portion of the large Kilnermium tribe, which had come out seal hunting on Albert Edward Bay that winter. Seven of them had died in the course of the winter. The Eskimo who told me this laid his hand on his chest and coughed to show us the cause; it must have been some lung disease. They were, if possible, more primitive than our Nechilli friends. They had to manage, to a greater extent than the Nechilli, with copper knives and copper needles, but their bows were better, and it was clear that they had more facilities for getting wood. Their dress was a little different from that of the Nechilli; the hoods had a blunt point at the back of the head. The waist band reached up high over the hips, but in compensation for this the "anoraks" were cut off shorter, being even more like dress coats than those of the Nechilli.

I bought a seal which the owner had brought home with him, and paid for it with a knife about two and a half inches long, which had been made from one of our ice-saw blades. It may have been rather mean, but if a transaction is honourable when both parties are satisfied, this one was so, because it would have been difficult to say which of us was the more satisfied with his bargain. I badly wanted the seal, and he, poor fellow, had never owned a steel knife in his life. Later we bought more blubber, the price for half a side of seal was a three-inch nail; but the price made even me blush. When the deal was concluded, and we had procured as much blubber as we could use, I gave the man a pair of surgical scissors as a present. Not only were he and his wife delighted, but the whole tribe rejoiced at the possession of such a treasure.

When we were resting in our sleeping bags we were constantly visited by some of the ladies, who brought us small mouthfuls of the cooked blubber from the front flippers of the bearded seal, which tasted something like pigs' trotters. At first we were innocent enough to believe that it was on account of our own personal attractions, but we subsequently discovered that it was some beads I had in a couple of match boxes that were the main attraction. I made them a present of about half a score; I had to be very economical with them. Gradually the mouthfuls became smaller, but in compensation they were more frequent; at last we had to send them off without remuneration; this had the desired effect, and we were not further disturbed.

For supper we cooked some blubber and meat of the seal I had bought. It was my first real meal of seal meat. I had certainly eaten seal meat before as a steak, but I had never tried it à la Eskimo, and without any Polar boasting I can say it was excellent. The meat tastes like mussels, and the blubber when fresh is just as delicate, though not so fat as pork. We had to be very careful not to put too much in the pot, for it slips down so easily that we run the risk of repeating the incident of the marrow soup at Cape Crozier.

On May 5th we remained where we were. Indeed, it would have been difficult to get our dogs to move, as they had been allowed to eat as much blubber as they liked, and that was no trifle after their long diet on reindeer meat. But it had proved rather too strong for them. We spent the day in cutting up the meat and getting our loads ready for the next day. We had to repack everything now that the greater part of our tin cases were wanted for the blubber. We removed the false runners from the sledges; we thought that the snow was now so wet that German-silver runners would glide more easily. We had sufficient food for one and a half months; we could not take any more because the sledges were full. We could have got more from the Eskimo, as they had been very successful with their catches, and had large stocks of blubber. The next night we slept just as well as the first. It was pleasant to lie down quietly with the knowledge that we were now to go over the level ice of Albert Edward Bay, and be able to make good headway again. There were probabilities of reaching new land, and these were pleasant thoughts to slumber on.

On May 6th we started off. Our friends were very contented with the few things they had obtained, a pair of scissors, a knife, a few nails, a match box of beads What riches! They would never forget the day. I wished we could have given them more, but we were not equipped for trading. Indeed, we had never expected to meet any Eskimo, but we determined if we encountered them on our return journey, to give them anything we could dispense with. With these good intentions we started out on the ice; the Eskimo stood looking after us as long as we could see them, probably discussing who we were and what our real object was. Possibly if they are left in peace in their poor land for another couple of centuries, two weird names, representing Ristvedt and myself may be handed down among their traditions. Our intention was to drive far enough into Albert Edward Bay to get inside the pack-ice. We had had enough of it, and even if we were obliged to proceed right to the end of the bay, that would be preferable to shortening the road; but, unfortunately, it would take much longer than following the direct route.

On May 7th we set our course direct for Cape Adelaide. We wanted to get on a height in order to have a view of the surroundings. We had long had to content ourselves with the poor view obtainable from a high hummock. We advanced over the southerly slope, which I fancy must be a paradise in the summer. There were already large patches quite free from snow, and some long grass was left there from the previous summer. The ground was certainly frozen quite hard and the grass withered, but we conjured up before our "snow-tired" eyes, fantastic pictures of green grass and rippling streams, of flowers and bilberries, of grazing reindeer, of hares tripping about, of lively ptarmigans, all bathed in the rays of the sun just as it might be on a lovely August day. An entire *fata Morgana* arose before the mind's eye, fortunately not altogether, however, for we descried some real ptarmigan and shot a brace.

On May 8th, towards noon, I noticed some jet black spots ahead of us. I stood looking at them for a while through the glasses, but as I could notice no movement I concluded they were stones. Certainly they were very isolated, apparently some distance from land, but I had long ago ceased to judge of distances in the winter light. Besides, we were accustomed to great surprises in the shape of long low projections of land spits, the existence of which we did not dream of till we drove over them and saw a stone here and there projecting from the snow. The sun had just come out, and we halted so that I might take the latitude at noon. While I fixed up the theodolite, Ristvedt took the glasses, and he arrived at a totally different conclusion from mine. The stones moved : they were seals—three huge ones lying basking in the sun. Their broad dark backs offered a good mark for the sun's rays. Anything white, such as snow and ice, throws back the rays of the sun, but black objects absorb them. There they lay sleeping, as seals do, for half a minute at a time. Every half minute they raise their heads, look round and sniff the air, then they let their heads droop again. So they go on without interruption, up and down, up and down. You must stealthily approach them when they are down, as then the seal neither sees nor hears anything, but it is all the more watchful when its head is raised. You must then lie as quiet as a mouse, concealed, if possible, behind some upstanding block of ice or other projection ; but, above all things, keep quiet. The least movement frightens the seal, and, however quiet, heavy, and sluggish it may have looked, it is gone like a flash of lightning. I was not therefore very confident in Ristvedt's success, but it was well worth his trying, while I attended to the observations, and accordingly off he went.

He had tied his dogs to the side of the sledge, and I had secured Silla close to the stern so that she could not pull without hurting her hind legs. Gjöa had a noose round her neck. These were the two worst I thought, and if they could not pull, the other ones would certainly keep quiet. But that is where I made a mistake. Just as I was lying down quietly to have a look at the sun through the instrument, which was standing upon a box at the side of the sledge, the box and the instrument received such a push that both toppled over in the snow in different directions, and away went the sledge. The dogs had heard the report of a gun, but I had not, and in spite of Silla's and Gjoa's efforts to hold on, off they went in Ristvedt's direction, straight along his track. But I was too quick for them, I got alongside and upset the sledge ; I presume the pack-ice had taught me that trick. In any case I succeeded in stopping them. Ristvedt's dogs, having to drag their sledge sideways, soon lost their inclination to run away. I then returned, set up the instrument again, and succeeded in getting the latitude. The instrument had fallen on such soft snow that it suffered no damage. After taking the latitude I drove up to Ristvedt. He had secured a bearded seal, 7 feet 6 inches long and six feet round at the fore-flippers—a regular mountain of flesh to look at. There were a number of small blocks of ice on the way, near the spot, and distributed so favourable that Ristvedt had succeeded in getting within range. Having selected one of the three seals lying round the hole in the ice, he aimed at its head and killed it. It is important to kill a seal outright; the shot must finish it off at once, otherwise even its last convulsion will cause its body to slide down through the hole in the ice, and then it is hopelessly lost. He had fired at the moment when it was looking up. It caused quite a thud on the ice as its head dropped; every muscle was relaxed. He ran up to plunge the harpoon into it; it was provided with a strap, and we carried it for any contingency. However, he preferred sending another bullet through its head at close quarters, to make quite sure of his booty. The size of the seal frightened him. Had he harpooned it, and there had been any life left in it, it might have regained consciousness for an instant and disappeared down the hole with the harpoon and line, possibly with Ristvedt himself, if he had attempted to prevent it. The hole was large enough for that, over two yards in circumference, with an even slope—a slide—on one side, where the seals could crawl up. When the shot was fired the other two disappeared from sight, apparently at the same instant. There must have been a little difference in time, however, as the hole was barely large enough for even one of the big creatures.

When starting on the morning of May 9th a fresh northerly breeze faced us. It was very cold, so much so that it was impossible to sit on the sledge, even with a full set of furs on, except for the few moments necessary for consulting the map. Whether the temperature was really 22° below zero Fahr., as we surmised, I cannot say. May be the high temperature of the past few days spoilt us. Now we had to find our way into the deep narrow bay charted by Rae, which cuts into the land to the north, from the northern coast of Albert Edward Bay. We wished to drive up through this bay, as I thought it could not be very far to the water beyond. Collinson has recorded a bay there. On driving for an hour and a-half we sighted land in the direction of our course. We approached it at a rapid pace, as the dogs had now shaken off the effect of their gourmandizing. It was no trifle they had managed to stow away, and at first it took some persuasion at the end of our whip to make them go at a reasonable speed. They could not understand why we should drive away from a spot where there was so much food. An Eskimo would have reasoned just the same, so presumably it was the most natural argument. It was civilisation that prompted our longing to make progress. This was a thought worth keeping clear in one's mind. Sometimes it required argument to convince ourselves that we were civilised men. The picture I had before me in the person of my fellow traveller, and the picture he had before him of his companion, as also our utensils and housekeeping arrangements, all the fat and dirt, petroleum, soot, reindeer hair and strips of skin, might easily make it doubtful. Our conversation about weather, wind, and hunting—*i.e.*, food—sleeping bags and dogs, and nothing else, would sometimes make us feel as though we were simply what we looked. That we throve exceedingly well on this: that we thought we had never eaten better in all our life, never slept so well: that in reality only sunshine, warmth, and food comprised all that we expected from our existence, might at times make us afraid of reflecting on what we were doing towards achieving our aim in life. But then we would again ponder on the trodden footmarks of two men and the narrow lines formed by the runners of our sledge on the virgin snow and the untrodden land beneath—the sledge tracks that ended at our tent, but which were to be continued to-morrow over the glittering expanse. After all, it was only a picture of strenuous endeavour. "To take life as one finds it" is called a virtue. Yes, of course, in a sense it is, but it comes very natural. Dogs have it and the Eskimo have it; men in whose hearts no such word as "Forward" is inscribed, rank no higher.

By degrees, as land began to loom out of the ice

haze, driving became more pleasant. It was something more to look at than the flat ice which previously surrounded us on every side. To be sure, there was no rocky land, but there were hills and slopes, some of the heights rising to about three hundred feet. We got up under the coast and entered a bay on the left, which we had thought of following. To all appearance it was the only opening on the coast. However, on reaching the lower end of it we found we were in the wrong street, and Ristvedt mounted a small hill to reconnoitre; thence he sighted the proper bay. He could distinguish the ice from the land by the ridge of ice extending all along the coast. This ice ridge is so formed that the sea practically freezes right down to the bottom in the shallow water near the beach. It is formed when the water is at its highest (spring tides). At ebb tide the sea recedes from under the ice, with the result that the latter, unable to support its own weight, drops down again on to the surface of the water, and then it breaks off near land, where it settles to the bottom as it cannot get away from the shore. The ice next the shore remains there, and the ice outside on the deeper water sinks a trifle more, thus forming a step, affording a means of ascending from the ice to the shore. The height of the step depends directly on the difference between the spring and ebb tides. Here the steps were a couple of feet high and enabled Ristvedt from that elevation to see that a narrow sound extended along the bottom of the valley. We had merely to drive across a narrow isthmus to get on the right road again.

It was my intention to drive to the end of the bay, but on getting down on the ice we perceived near the other coast an island, with a very peculiar erection, like a cairn, at the highest point. This, of course, had to be investigated more closely. The Eskimo do not build such cairns, they merely place single stones on the top of one another This one, however, looked quite monumental. We drove across the bay and pitched our camp near the foreshore of the island. Sure enough, it was a cairn, although not quite so high as it appeared at first sight. Across the top of the island there was a ridge about fifteen feet high. At the western end of this ridge there was another pyramidal elevation, and the cairn, consisting of large slabs of limestone, was built on the summit. We pulled down the cairn. I have always had an objection to this work of demolition. Cairns, miserable stoneheaps though they be, are signs of human beings, human work, in the midst of the wild deserts. But that is not all. One's hesitation to take it down is due to one's veneration for the men who have been there before. The cairns meant something to them, just the same as those we erect have a meaning for ourselves. Some difficulty surmounted, some step forward towards a goal. They leave a trace of our wanderings that is to endure for centuries, when the snow has long since melted from under the sledge tracks and when our names have disappeared like the melting snow. They are a trophy of victory impressed upon land, won from darkness, from the spirits of evil. But the cairn had to come down, we had to see whether it contained anything, perhaps a message from our brave precursor, Rae. We found nothing, however. We thereupon descended to our camp. On our way down we saw a hare, but tried in vain to use Gjöa as a sporting dog. She did not seem to take the slightest interest in the subject, and when the hare took to its heels across the ice, we abandoned the idea of that little luxury.

On May 10th we placed a depôt beneath Rae's stones. All we had left of the dog's food, prepared by ourselves—about 1 cwt.—was deposited there. We proposed using it during the difficult drive over Victoria ice. We were kept prisoners for two days. During that time we discussed what we should do in case the ice on Victoria Strait broke up before we reached home. It was just as well to talk about it some time in advance. However, now that we had made a regular start on the new land, and there was every chance that we could procure the necessary food ourselves, we were tempted to prolong our survey into the spring.

We left on the 13th. We had prepared the sledges on the previous night, and in the morning we were awakened by the ptarmigan outside our tent. It sounded very much like summer, but it looked anything but summer-like. It was a regular snow-storm, but, having the wind behind us, we decided on starting. We passed over a few isolated slopes on the way, and from the summit of one of them, Alice Hill, we sighted pack ice in a northerly direction. We had another eight miles to drive, however, before we got there. I may mention that there was no real bay where Collinson had charted one; he had drawn his map of this section at a distance. The point "Collinson's Farthest" was situated some distance to the east of the spot where we reached water. From there he had taken the isolated slopes for islands and capes, and the low ground between them for water. The low stretches of lakes and swampy ground we passed could not possibly be distinguished at a distance. When a slope of this description extends out to sea, and when there is little difference between the ebb and spring-tide, as is the case at that spot, one cannot tell where the land ends and where the sea begins, except when driving across the boundary line. We reached the beach close to a small headland and thereupon branched off on the ice in a north-westerly direction. We soon sighted land, which, however, only consisted of a group of islets and skerries. Looking due west we saw a higher point inland near the coast of the mainland. We drove over and remained there the night. Our observations during the day were not of much importance. The snow-storm prevented us from distinguishing anything clearly.

May 15th was a very cold day. The temperature was down to 22° Fahr. below zero. It blew a little, but the weather was clear enough for us to start. Before leaving camp we fixed up one of our false runners on the top of the hillock to have a mark for the telescope of the plane table. I had previously noticed how quickly one loses sight of cairns of snow, even when built to the height of a man. It is impossible to distinguish them at a distance of a couple of miles, even in clear weather. The false runner was intended to help us to recognise our hillock at a distance. After driving for a couple of hours we reached a low clear slope at the end of a headland, Cape Kofoed Hansen. We erected a cairn there and drove across a bay—Homan's Bay—the second headland of which we sighted to the N.N.W. Midway on the bay we stopped to survey. There were two points from which I could not see the far end of the bay, but when doing charting of this kind one has not always time to enter into little details. The main thing is to get the coast line fairly outlined as far as possible.

May 16th was another idle day, thanks to the frightful weather. It gave us an opportunity of looking after our clothes properly, but that was its only redeeming feature, for we had no time to be really idle. We were getting short of fuel. A couple of unfortunate upsets had lost us some petroleum, and by gauging our stock on May 13th we discovered that we had only one and three quarters gallons left. We must therefore economise.

May 17th was Norwegian Independence Day. We began the day by breaking open a small box which Lindström had presented to us for this occasion on leaving. We had for some time regarded it with curious glances and tried to make each other believe that it was silly to be dragging it about on the sledge: that it would be much better to open it at once. But, fortunately, it always happened that when one was weak the other was strong, and when one was for opening the box the other descanted on the enormity of so doing before the appointed time. The result was that the box remained intact. It contained a fish-pudding, two tins of milk, some citron-pudding powder, wheaten bread, and six cigars, the whole constituting a very acceptable present

Peder Ristvedt.

AT VICTORIA LAND.

for gentlemen in our present position. We each lighted a cigar forthwith, and when we had finished smoking we struck camp. The weather was not particularly fine, and we found it difficult to make progress from the very start. The snowstorm on the previous day had caused a quantity of loose snow to accumulate between the big drifts around the mountain of ice, and we had almost to swim through it, both men and dogs, to reach better ground; but, in fact, it was bad enough all the way.

Sometimes we would drive a little way in on the coast and sometimes on the ice again. It was equally bad on both, and we waded through snow knee-deep. We reached a headland, Cape Christian Mikkelsen. To the north of this a deep fjord runs inland. Later on, the Captain named it Denmark Fjord. At the mouth of this fjord there was an island falling off steeply to the south. The island was probably about 180 feet high, considerably higher than the land along the sides of the fjord, seen from it summit. Remembering what day it was, I asked Ristvedt the name of the most important person who took part in forming the Norwegian Constitution. "Falsen!" he said, and so we named the island.. We then drove a little farther on towards some high land which we noticed on the north side of the fjord, but at 5 o'clock we halted for the day. It was the first time for many a day that we had halted to fix camp simply because we were tired and too fagged to drive any further. After we had had our dinner, Ristvedt made a citron-pudding in our chocolate-pot. Of course we partook of it too freely.

The weather was so fine on the morning of May 18th that we were in good hopes it had now set in in earnest. I did some surveying, and then we proceeded towards the same high land we had been making for on the previous day. Soon, however, everything disappeared in a haze, and we had to steer our course as best we could from one section of the main ice to another. It is not very convenient to stop too often to verify the course by means of the compass. In such heavy sledging as we had to contend with there, the dogs seemed to become stiff-legged by waiting, instead of deriving benefit from the little rest. I thought I was now sufficiently experienced not to mistake land for water, however much snow there might be, but that was a fallacy. While driving merrily along in the belief that I was on ice all the time, a stone suddenly protruded from the snow. Evidently there was no water here. Soon after I noticed a small piece of broken ice standing up on edge. Certainly this could not be land. But the transition had quite escaped my attention. I charted the whole as a ridge of skerries and islets round the north headland of the fjord, Cape Peter Anker. I think this will turn out to be right, when checked in the summer.

We camped at Cape Nygaard. We had seen from Cape Anker that the land ended here; it was a ridge extending from Mount Dirckinck Holmfeld, out towards Cape Anker. Ahead of us we could see nothing. Sky, land, and ice intermingled in one indefinite grey haze. The clouds looked threatening. A large ring surrounded the sun in the afternoon. Cape Anker lying behind us was only recognised from the fact that the shallows off the land at that point had stopped the larger ice, so that the field beyond was smooth. Only Mount Dirckinck Holmfeld, with its gradients, was illuminated by a strong silvery light. Everything pointed to bad weather. We were being dosed with it and no mistake.

On May 20th we had for the first time a temperature

of about 32° Fahr. Sleet, a mixture of snow and rain, was falling when we awoke. The flakes falling upon the canvas of our tent were large, and every now and then they melted, leaving a wet sparkling spot on the tent. Yet the weather was improving. The wind cleared the atmosphere later in the day. However, it was out of the question to think of moving before night. The wet snow would have formed big lumps under the runners of our sledge. We did not make a start till 9 P.M. The temperature had fallen during the afternoon, with the result that a crust of ice had formed on the snow, and this made excellent going. We spun along in magnificent weather. Heavy clouds were forming all round the firmament, the sun throwing streams of light and dark rays down upon the ice. It was beautiful, yet there was something uncanny, uncertain, and tempestuous about the illumination. We dared not hope it would keep fine for long—neither did it. The sun set at 11 o'clock behind a wall of bluish-black clouds. We had been out some distance from land on a portion of the main ice taking coast bearings, but now we set our course direct for a high cape. On our way we passed a long, low headland—Point Dietrichsen—just as it started snowing. We built a cairn there, hoping that this was only a squall of short duration, and that it would be over by the time we had finished our work; but it grew worse and worse, so we drove out into the haze of snow and were fortunate enough to get direct to the high land. It was an isolated hill, which I at first took for an island. It turned out later that it was a final spur extending from the slopes round Mount Dirckinck Holmfeld. The cape, the most prominent point on the coast, was christened Cape Sverdrup.

At 2 A.M. on May 21st we encamped. The drive during the night had made us so sleepy that we cooked a little pemmican for our meal. We did not care to wait the hour or so that would be required for preparing meat from our "fodder-box." When we woke again, quite eleven hours later, we were very hungry. It seemed as if the meal of pemmican had not been substantial enough. We had only used the regulation rations, 14 ounces. This allowance included some chocolate, but we had not been able to prepare it on account of the scarcity of fuel. We quite recognised that the ration of pemmican for the day's chief meal, $1\frac{4}{5}$ lbs. of food, could on no account be reduced. We now treated ourselves to an extra cup of chocolate.

It was rather a depressing thought that we should soon have to turn back. The bad weather had only one advantage—we were not tempted too far away from our base. Yet it was a pity to be reduced to finding consolation in a line of argument that ought to be foreign to a fit and energetic Arctic traveller.

We escaped with one day of idleness at Cape Sverdrup. On the 22nd we got away. With a fresh breeze straight in our faces we proceeded over the broad flat bay, Norway Bay, to the north of Cape Sverdrup. Before our departure I went up to Cape Sverdrup to take observations, and came upon an old tent ring there. On the very top there was a huge boulder, on which I placed a long stone taken from the tent ring. This kind of cairn is visible from a much greater distance, and is more substantial than the ordinary ones. The land along the bay was quite flat, so also was the northern headland itself, Point Isachsen. When we reached this, I perceived Ristvedt running after me, and stopped for him. He asked me to lend him the glasses, and I thought there must be something the matter. I glanced over the ice, however, and saw nothing, so I imagined he had made a mistake. But he had seen enough—"Bear," he whispered. His lips formed the words rather than uttered them. His sporting eagerness had got hold of him; he feared to scare his quarry, but this was rather unnecessary. There was a bear, no doubt, but it was a long way off. It was standing out on a large plateau of smooth ice, extending outwards from Point Isachsen and meeting the pack-ice quite near the horizon. Its head poised on a wonderfully long neck, drooped down towards the ice; its legs resembled four short columns. There was nothing very terrifying about it in that position; it was probably standing there asleep. Its yellowish skin was much the same colour as a block of dirty pack-ice; it was almost impossible to distinguish it from the surrounding blocks of ice. Suddenly it turned its head round towards us, and then we could see its black snout quite distinctly. No other object in this kingdom of the Snow Queen, no stone, no bare spot, no dark shadow is so black as the snout of a Polar bear. It cannot be mistaken even miles away. It looked in our direction, probably it had heard some of our dogs barking ; however, it soon turned it head away again and continued dozing.

Ristvedt took Silla and went off towards the bear. When he got within a suitable distance, he let the bitch loose, after showing her the direction. Away she went, slowly at first, but suddenly she seemed to catch sight of the bear, and then forward she darted, like a black streak over the ice. I loosened more dogs. They had by this time learnt what it meant when Ristvedt went off with his rifle. Like a flash they followed his track and passed him, as he stumbled along over the snow, away after Silla. I noticed through the glasses how the bear again slowly turned its head towards the sledge. The dogs never barked ; it was possibly by the merest coincidence that it turned its head. But the sight that then met its eyes soon made it lively. It wheeled right round and lifted its head, stood its ground, but only for an instant. By instinct, or from experience, it must know the wolves that can chase even a bear till it drops from sheer exhaustion, though each one of them be such a miserably small creature compared to its majestic self. It now perceived a leash of six—seven—eight coming towards it at a breakneck speed, black spots dotted over the ice. Life was in jeopardy ; the peacefulness of the desert was disturbed. Up it bounded, all its four legs off the ice at once, and wheeled round. Then off at a gallop towards the pack-ice, full pelt, to save its life.

Silla was too quick for Bruin. Just before the bear could reach the edge of the life-saving pack-ice, in the maze of which—where it was quite at home and within a yard or so of safety—the dogs could not have kept up with it, it was overtaken. Silla bounded forward and fastened her teeth in its tail. The bear had to stop and throw itself round to shake off the enemy, and round it spun, at such a rate that Silla had to let go her grip and was sent flying on the ice. Scarcely had the bear turned to resume its flight than Silla was on it again. This was repeated several times, but meanwhile the other dogs had caught up to the fugitive, and then there was no hope for the bear. The dogs surrounded it, and no matter to which side it turned, some dog or other was at its heels. Through the glasses I could see it spinning round, bounding off the ice like a rubber ball, at a speed one would not have credited in such a clumsy creature. Its rage increased. Then Ristvedt got up and fired. The bear was wounded, but not mortally; it merely rose on its hind-legs and fought the air with its fore-paws. The dogs closed in on it. Silla, in her fury, sprang right at the face of the bear, and received a blow from its broad paw. It must already have been somewhat weakened, otherwise it would have killed the plucky bitch outright. As it was, she took several minutes to recover; we had given up hopes of her; her eagerness for bear-hunting was knocked out of her now. The bear died in the belief that it had killed the enemy who had first overtaken it and stayed its flight from death. It was shot through the head by Ristvedt's second ball, fired at a moment when the dogs were timidly holding back after Silla had been struck, thereby enabling him to fire without hitting one of them. It was a lean young she-bear, with no trace of food in her inside, and a curious hide-like layer of adipose tissue, almost devoid of fat. We had bear-flesh for supper. The soup was good, but the flesh was very coarse and stringy. Besides, it was so lean that we were soon hungry again, notwithstanding the ample helpings. To bring the soup up to the proper standard, we ought to have put a good-sized lump of seal blubber into the pot.

The same night we had an illustration of the difficulty of gauging the size of things under certain conditions of light. We had turned in when Ristvedt thought he would have another peep through the little hole in the door to see if everything was right, when he saw another bear. We opened the door in double-quick time and sure enough we saw a dirty yellowish object running away on the ice. Per, Bay, and Silla were after it full tilt. Contrary to our custom, we had not tied the dogs up as the bear we had killed was too large to carry away with us, and we did not mind the dogs having a feed during the night. In any case they would find it preferable to sailcloth and tarred rope. Now they were after the bear in full chase. They overtook it and then began to waltz round it in the same fashion as we had already seen earlier in the day. The waltz suddenly ceased, and the dogs came running back towards us, Per carrying the "bear" in its mouth. I don't know whether it was the light alone which accounted for this, or defective observation during our hurried preparations, or the blind sporting fever that undeniably seizes one when confronted with such big game as King Bruin, but we felt rather crestfallen at the finish when the "bear" turned out to be nothing more than a white fox. Full of the pride of victory, Per arrived with the fox in its mouth and laid it down. Next day it was consigned to the cooking pot. Fox-flesh has a peculiar flavour, and the odour reminds one of the smell of a wild beast cage; apart from this, it is, in the matter of fibre, the best meat procurable during the winter. Bear is stringy, ptarmigan is tough, and reindeer like firewood, but the fox keeps himself in good condition all through the winter, so that the fat is suitably distributed, and the meat tender. There is not much feeding on him, as he is not much bigger than a good-sized rabbit.

It was a trifle troublesome to get the dogs to move next morning, May 23rd, just as it had been on the other occasions when they had been allowed to eat their fill. They groaned and panted in the hot sunshine. But as it was nothing more than the headache following the debauch, we did not hesitate to use the whip to urge them on. We never used it as a rule, and, of course, we were more loth to do so when the dogs had been on short rations for some time. We made good progress to-day. The temperature had been 14° Fahr., but being a beautiful calm day the sun had had sufficient effect on the snow to cake it under the runners. The dogs suffered from the heat, so we decided to drive during the night in future.

On May 25th it was fine all the day, and we expected the best results from our latest march; but we had hardly made a start in the evening before a bank of clouds rose very rapidly over the northern horizon. It was just as though the sun fell from the clouds, so quickly did they gather. Before we could count ten the beautiful evening, with the glowing midnight sun and golden purple clouds, with all its other glories, had changed into a cold, clammy, dismal, dark autumn night. We had to get along, however, as we meant to cover another twenty miles. Now and then we adjusted our course by a pocket compass. To judge from appearances we were on a bay; the ice was quite smooth. Soon after midnight we reached land, and then the fog lifted a little. To our astonishment we saw land on all sides of us. We had driven into a bay that was completely land-locked. We called it "Greely Harbour." This would be a splendid winter-harbour; it is the only good one on the coast. There was plenty of deep water as we could see by two or three large masses of sea-ice that had found their way into the bay; they must have been submerged quite five fathoms. The land on the west side of the bay was high, and that on the east side also. Between two slopes there was a very low narrow isthmus, over which we drove out on to the sea-ice again, and thence we followed the coast northwards. Unfortunately it did not clear up after all. The fog again enveloped everything, gloomy, cold and clammy. This damped my courage. Why struggle to advance, when we could not even see land, and the most we could have done was only a matter of another couple of miles. So we halted and went into camp off a low rocky headland, very much like all the others we had passed. This we called "Hansen's Farthest."

Who has not at some time in his life stood overpowered with the conviction, "Thus far and no farther." Those who have, know how depressing it is. We had long recognised that we could not reach our goal. When we were labouring and toiling over the pack ice out on Victoria Strait we had more than once remarked that we would content ourselves with one or two days' journey along new land; but although we had done a good deal more than that, still we were rather disappointed at not having reached Glenelg Bay. We had to leave the hundred miles separating us from it, in their untrodden virgin whiteness. We did not conquer them.

On that day when I closed my journal with the words: "Thus far and no farther," I felt almost as if the whole of our labour had been wasted. When "Forward" is your watchword your muscles are tense, your heart beats, the blood courses through your veins, your head is erect, and your form upright. But "Backward" seems to turn you at once into a decrepit old man. It was some days before we could again reconcile ourselves to the words of Björnson's poem :

> " Loft dit Hoved, du raske Gut !
> Om et Halb eller to blev brudt,
> Blinker et nyt i dit Oje."

We prepared the following document to be left in a cairn :—

" A sledge expedition from the Norwegian 'Gjöa' Expedition reached this point on May 26th, 1905, and named it 'Cape Nansen.

" It is situated 72° 2′ N., 104° 45′ W. (Greenwich).

" The coast appears to continue in a N.W. direction.

" We are returning to the ship on this date.

"Cape Nansen, May 26th, 1905.

" GODFRED HANSEN.
" PER RISTVEDT."

Turning "back," as I have said, makes one feel old and broken up, but turning "homeward" is quite a different thing, We were now going home, home in real earnest. Cape Nansen was our farthest point ; the return journey would not terminate in Gjöahavn, it would continue in the "Gjöa" as soon as the ice opened—then onward, homeward.

At Cape Nansen we secured a bear. It was on the 26th, in the evening, just as I was preparing to strike camp. Ristvedt looked out through the door of the tent. But, instead of the sledges and the dogs forming a crescent round the tent, and the fading outline of some hummocks, and beyond grey nothingness, snow and ice melting into a haze : instead of all this, he saw a bear standing some ten feet from the entrance to the tent. The rifle was always lying in the tent, loaded. In an instant Ristvedt picked it up and fired. The bear fell, but got up again and hobbled away on the pack ice. The blood was running from its throat. We rushed out, barefooted as we were, to loose the dogs. But our hurry was unnecessary. The bear only got some thirty paces away, and then rolled over stone dead.

May 27th.—The land north of Cape Nansen was again quite flat. To the south I could see the high land surrounding Greely Harbour. The most prominent, however, was Mount Ovidias towering high above the perfectly level plain extending for miles round its base. It is the most easily recognised point on the entire coast. I then returned to the tent, and our homeward journey began.

At Cape Anker we turned into Denmark Fjord, and passed the night on Clöette Island, some ten miles up the fjord. Although the weather was clear, we could not as yet see the far end of the fjord, and we made up our minds to proceed some distance further up next day. According to the map I had sketched on our outward journey, there should be a neck of land farther in, about four miles wide, between Homan's Bay and Denmark Fjord. I wanted to drive over there to check the work, which had been carried out only by means of bearings taken from the outside.

It was surprising to notice, as we proceeded along the coast, how quickly the fine weather altered the appearance of the land. On June 1st, the sun, which was now visible in the sky throughout the day, was shining with its full force on the snow, melting it, and laying bare the black earth beneath. As yet there were no running streams apparent, but the snow-heaps continued to diminish every day. In places that were thinly covered it disappeared altogether, so that the hill crests became almost black. We soon began camping on the ground, as we found that the warmth of our bodies penetrated through our sleeping bags and the tent floor and melted the snow, so that the bags became damp. Camping on bare ground was therefore preferable. Even if the bed were not so smooth as it might have been, we were not so particular as the "Princess" in the story of the "Pea," and we did not mind putting up with a small stone or so in our beds. We saw hares while proceeding down along the coast, often three to four at a time. Some of them found their way into our pot. It was difficult, however, to get within gun range, and we dared not be too reckless with our ammunition. It was meant for bigger game. Off Cape Kofoed Hansen, we shot another bear.

On June 5th we passed "Rae's Cairn" Island and found our depôt all right. The lemmings had had a few mouthfuls, but not more than we could very well spare. This is, I may say, an animal for which I have a certain amount of respect. Should anyone wound its spirit of independence by crossing its path, it resolutely rises on its hind legs, with its back against a stone if possible. Sitting on its hind legs it fights with its fore paws in the air, for all the world like a bear, ready to sell its life as dearly as possible. As you stand in front of it, towering into the skies, while the tiny creature only reaches to your ankle, you can scarcely help laughing at such a curious exhibition of courage. Nevertheless, it commands respect. We reached our depôt of bearded seal on the ice next morning. It took us some time to discover it.

On June 11th, just after midnight, we drove on, along Tayler Island. We found traces of two Eskimo sledges going south, and we followed them, hoping to overtake the Eskimo. For some reason or other, however, they must have been in a hurry to get south. Possibly they were to meet kinsmen and had postponed the journey as long as possible, hoping that we would return. They had made no halt for the night all the way down to Dehaven Point. From there we crossed straight over to Lind Island, which was prominently visible to the south. We halted for the day in the middle of the sound.

On June 14th we tackled the pack-ice. I had thought of it with some misgivings, in view of the possible scars and inevitable exertions. We were, however, let off easily. In two days we reached the "Land seen by Rae." On the first day in the pack-ice we caught a seal. It was most welcome, as the bearded seal blubber had become somewhat rancid. We had, in fact, seen many seals since we began camping in the open. On smooth floes of somewhat considerable extent one could be certain of seeing seals that had come up to bask in the sun. They were very shy, and it was only possible to get within shooting range in places where heaps of pack-ice had accumulated conveniently.

We reached land early in the morning of the 15th. It was only an islet, but we sighted land, large and small islands, to the north, south, and east. It was quite summer on the islet, hardly any snow, fresh green moss, ptarmigan and eider-ducks. To us it seemed quite a regular little paradise, and we named it Princess Ingeborg's Island. I determined its longitude and latitude. The charting of the other islands was done in a somewhat perfunctory manner; as we were, in fact, in somewhat reduced circumstances. We cooked with blubber, as we wanted to save up the remaining half-pint of petroleum lest we should get such bad weather some day as to make cooking outdoors impossible. Our supply of bread was exhausted, and we had only two rations of chocolate left. They would soon be anxiously awaiting us at home. All this caused me to hasten our progress.

We made a day's journey of twenty-one miles south, passing between a great number of islands and islets. For the last two or three miles we drove over a Strait—Markham Strait—and I had an idea that there must be sufficient depth for the "Gjöa." We reached Bryde's

Island, south of the Strait, from the top of which I sighted several islands to the south, and at the farthest point, probably about fifteen milės off, high hilly land, apparently an island of some extent. I had an idea of examining the group of islands farther southwards, and had started to do so, but thinking that the group extended right to the mainland, I concluded that the task would take too long, considering the advanced season, and I therefore altered our course northwards again, to the east of the islands.

We named the group south of Bryde's Island "Nordenskjöld's Islands." The group north of Markham Strait was named "Royal Geographical Society's Islands," and the most prominent points of the islands were given English names. This seemed to us the most appropriate, as the land was first sighted by an Englishman.

On our way along the east coast we lost a dog. Ristvedt had taken it out of the team, as it was impossible to make it work, and it only caused trouble among the other dogs. He was a reddish coated dog, with short legs, answering the name of "Inagsayak." He had been lazy all the time, and, as the dogs had for some time been fed on blubber mixture, it had grown rather fat. It followed behind the sledge for a time, but at last even this was too much for it; it lay down on the ice and there it remained. We saw no more of it. We fully expected it to turn up at our tent in the evening, but it did not. It must have died there. Our consciences pricked us a little at first, but there was, apparently, nothing the matter with it when unharnessed.

We therefore consoled ourselves with the thought thа if it had died, its death was due to its own laziness.

On June 18th we started off across Alexandra Strai The snow in the pack ice had altered in a deplorabl manner. The crust of ice formed during the night wa not thick enough to carry us and the dogs, but, fortu nately, the sledges floated on the surface, and we mad some progress after all. At Cape Crozier we found ou reindeer-meat, our petroleum and our chocolate, all i good order. Our privations were at an end. We no had the smooth ice of Simpson Strait to drive on, an made quick marches along the coast.

One thing, however, was very embarrassing at th late season: our dogs got bad feet. The snow was o the ice, and the ice-water had rendered the surface roug and uneven, with numberless small perpendicular icicle which injured the dogs' feet. There was blood on th track after them, and one after the other they had to b unharnessed—first Mylius, then Gjöa, then Silla. W just managed to get them to follow behind our sledge If travelling is unavoidable when the ice is in an advance state of thaw, the dogs should be provided with suitabl foot-gear, otherwise the work is too much for then After a time, as we advanced, the features of the landscap gradually became more familiar. We had often visite the coast as far as the narrow part of the Strait. Afte passing Todd Island, and turning into Peterson's Ba we began to feel quite at home. Our last camp was a Svartheia (the Black Mountain).

On the morning of June 25th we started on our last ten miles. Our sledge-flag, which had become somewhat tattered, floated over the sledge from the end of a ski. We intended to be seen by those on board at the earliest moment, and they soon saw it, as they were anxiously looking out for us. At 7 A.M. the flag was hoisted on board. Lund was on the look-out that morning, and had seen us immediately. We learned afterwards how many times they had looked out for us on the ice, but in vain, nothing but the level ice and Todd Island in the distance dancing up and down in the haze; but finally we arrived. We entered the mouth of the harbour at 8 o'clock and soon reached the vessel. The dogs had suddenly recognised the place, and realised that they were going to have a thorough rest. A man came down from the vessel, advancing towards us with long strides. It was the Commander. "God dag og velkommen" ("Good-day and welcome"), said he, and welcome we were, that was evident.

The journey was at an end. It had been a trying trip; you have to keep wide awake when travelling in the deserts. Ever such a little blunder and you may lose your life. A mistake means death. However, it is a manly life; you feel free when out there, where will is law even though it is hard, for the road is strewn with difficulties. One makes acquaintance with hunger, cold, wet and fatigue. The fare is frugal. You have to say good-bye to cleanliness, when every drop of water has to be produced at the expense of the most precious of all your possessions, fuel. However, on you go, and every mile covered seems another victory. And life : *La vie n'est pas un plaiseur ni une douleur, mais une affaire grave, dont nous sommes chargée, et qu'il faut conduire et terminer à notre honneur.*"

We had achieved this ; we had charted another stretch of coast on the blank part of the Northern Hemisphere ; we had caused new land to be trodden by the foot of man, and had made this land, its geology, its physical conditions, and its geography known. During our journey we had covered 800 miles.

Conclusion.

As I sat out there on my sledge, without any guide-post ahead of me, the runners of my sledge making the first tracks through those fields of snow, I often thought it would be a good subject to write about on my return home. It seemed to me that our journey was not without importance. Though the coast along which we drove was stern, stormy, foggy, and ice-bound both in summer and winter, though the land we wrested from the realm of darkness, and mapped out on our chart was barren and stony, shorn of natural beauty, useless to mankind, yet it seemed to me that the infinite wastes gave birth to conceptions of greatness, beauty, and goodness. This was to be my theme. I desired so to write that those who would read might enrich their ideas and gain some impressions, at least, of the Stupendous, such as were conveyed to me in those pathless regions, where God's sun or the bright stars alone point the way. Now that I have come to the end of my task, I realise how little I have been able to offer, because the thoughts that to me were overwhelming, are such as find expression in the soul rather than on the lips. If, however, I have, to some extent, succeeded in telling the story of two men and twelve dogs wading through snow, crawling over ice, resting in the lonely tent, exposed to the winds; if I have only once succeeded in faintly picturing the impressions produced by what is seen out there in the endless expanse, in storm and in sunshine, I shall have done something more than add to the chart a few miles of land north of "Collinson's Farthest."

Addendum by Captain Roald Amundsen.

I desire to express my most respectful and hearty thanks to all those who have lent their kind support to the "Gjöa" Expedition by contributions of money, goods, or presents, or by undertaking guarantees.

	Kr.	Öre.
H.M. King Haakon VII	10,000	00
H.M. King Oscar II	10,000	00
The Norwegian Government	40,000	00
Mr. N. A. Stang, Merchant	10,000	00
Mr. P. M. Anker, Landed Proprietor . .	10,000	00
Mr. Mads Wiel, Merchant	5,000	00
His Excellency Fridtjof Nansen, Minister of State	5,000	00
The Nansen Fund	5,000	00
Mrs. Olava Christiansen	3,000	00
Mr. W. Nygaard, Publisher	2,500	00
Mr. Haaken Mathiesen, Chamberlain . .	2,000	00
Anonymous	2,000	00
Royal Geographical Society	1,800	00
Mr. K. R. Berg, Manufacturer	1,000	00
Mr. Kristen Irgens, B.A.	1,000	00
Mr. Gunnar Knudsen, President of the Storting .	1,000	00
The Östlandske Petroleums Co.	1,000	00
Mr. C. H. Homan, Barrister	1,000	00
Mr. Ths. Fearnley, Master of the Royal Hunt .	1,000	00
Mr. T. H. Schjelderup, Merchant . . .	1,000	00
Mr. Johs. G. Heftye, Telegraph Director . .	1,000	00
Mrs. Evenstad	1,000	00
Christiania Söforsikringsselskab	1,000	00
Mr. Hans Kjaer, Merchant	1,000	00
Mr. Carl Lövenskiold, Minister of State . .	1,000	00
Mr. M. W. Stand, Consul	1,000	00
Mr. Axel Heiberg, Consul	1,000	00
Carried forward . .	120,300	00

	Kr.	Öre.
Brought forward . .	120,300	00
Mr. Joh. Thorne, Councillor of State . .	1,000	00
Mr. Ellef Ringnes, Brewery Proprietor . .	1,000	00
Messrs. O. Mustad and Son	1,000	00
Mr. Chr. Schou, Manufacturer . . .	1,000	00
Mr. Thv. Meyer, Merchant . . .	1,000	00
Miss Harriet Wedel Jarlsberg	1,000	00
Mr. J. C. Juel, Merchant	1,000	00
Mr. H. F. Dessen, Merchant, London . .	1,000	00
Mr. Th. Fagelund, Shipowner, London . .	1,000	00
Mr. Jacob Hessler, Merchant, West Hartlepool	1,000	00
Mr. J. Jörgensen, Merchant, London . .	1,000	00
"A Friend," through Mr. Fagelund, Shipowner	500	00
Mr. J. C. Pharo, Merchant, London . .	500	00
Mr. Johan. Anker, Engineer	500	00
Mr. J. Holst, Merchant, through Mr. Fagelund, Shipowner	456	30
Mr. J. W. Constantin Schröter, Cardiff . .	200	00
Mr. L. W. Longstaff	180	00
Union Internationale, Antwerp, through Kjeld Stud and Co.	100	00
Through Mr. Arvid Bergvall—		
Kgl. Oct. Sö-Assurance Kompagnie . .	100	00
The British Dominion Marine Insurance Co.	100	00
Allgemeine Seeversicherungs Gesellschaft .	50	00
Total	133,986	30
About	£7,440	

Contributions in the shape of Goods and Instruments.

Provisions and instruments from Mr. Axel Heiberg, Consul, and Messrs. Ringnes Brothers.

Instruments from the Meteorological Institute and the Astronomical Observatory, also from the International Central Laboratory for the Study of the Sea.

Chocolate from Messrs. Brödrena Cloetta.

Tobacco from Mr. J. L. Tiedemann (Mr. Joh. H. Andresen), and Johs. N. With's Tobaksfabrik A. G.

Matches from the Nitedals Tændstiksfabrik.

Drugs from Mr. Ths. O. Alstad, Chemist, and Mr. Tillier, Chemist.

Medicine Chest Outfit from Messrs. Nyegaard and Co.

Surgical Instruments and Dressings from Mr. Christian Falchenberg.

Instruments from Mr. Fr. Aug. Michelet.

Gunpowder from the Nitedal Powder Factory.

Books from Messrs. H. Aschehoug and Co., Mr. Jacob Dybwad, and Messrs. Feilberg and Landmark.

Christmas and Birthday presents from relatives, friends, and acquaintances.

Loan and Guarantees.

Bank Loan through Mr. O. Ditlev-Simonsen, Shipowner, with the following Joint Guarantors :—

Mr. Einar Björnson, Director.
Mr. Hans. Br. Blehr, Shipbroker.
Mr. Johan Bryde, Shipowner.
Mr. Ludwig Castberg, Shipowner.
Mr. Joachim Greig, Shipowner.
Mr. Ivor Klaveness, Shipowner.
Mr. Fred Olsen, Shipowner.
Mr. Th. Pedersen, Shipowner.
Mr. Hj. Siegwarth, Shipowner.
Mr. D. Ditlev-Simonsen, Shipowner.
Mr. Chr. P. Staubo, Shipowner.

Supplement.

Cash loan, free of interest, from Jens Amundsen, Shipowner of Fredrikshald, 1,000 kr.

Loan on bills through Dr. Fridtjof Nansen, Minister :—

	Kr.
Dr. Fridtjof Nansen, Minister	6,000
Mrs. Evenstad, Widow	4,000
Mr. Johs. Heftye, Banker	3,000
Captain Scott-Hansen, R.N. . . .	1,000
Total	14,000

Guarantee for payment of debt, through Mr. C. H. Homan, Barrister, to the amount of 2,120 kr. 65 öre.

THE END.

INDEX

Index

Index

2 B 2

Index

Index

Index

Index

Index

Index

Index

Index

Index

Index

Index

Index

Index

Index

Index

Index

LONDON:
HARRISON AND SONS, ST. MARTIN'S LANE,
PRINTERS IN ORDINARY TO HIS MAJESTY.

14947279R00248

Made in the USA
Lexington, KY
01 May 2012